PROBLEMS IN PHYSICAL CHEMISTRY

PROBLEMS IN PHYSICAL CHEMISTRY

R.L. Laksh

RAJAT PUBLICATIONS
New Delhi

Problems in Physical Chemistry
by *R.L. Laksh*

This edition is published by Rajat Publications
4740/23, Ansari Road Daryaganj,
New Delhi - 110002

ISBN: 978-81-7880-058-5

Edition 2026

Printed and bound in India

RAJAT
PUBLICATIONS

4740/23, Ansari Road, Daryaganj,
New Delhi- 110002 (India)
Mob:- +91-9811484973, +91-9811105389
E-mail: rajat_publication@yahoo.com

Contents

1

Kinetic Theory of Gases

Q. 1 : The bar is unit of pressure defined as 1.00×10^6 dynes/cm^2, or 1.00×10^5 N/m^2. What is the equivalent of the bar in units of mm Hg ?

Solution : This is a matter of scale conversions. In a problem dealing with pressure, it is reasonable to use a scale of One Atmospheric Pressure = 1.013×10^5 N/m^2 = 760 mm Hg for conversions. Note the convenience in appropriate scale choice. If an inapropriate scale had been chosen, another conversion to some other scale would be needed before converting to the desired scale. Since 1.013×10^5 N/m^2 is equivalent to 760 mm Hg, therefore 1.00×10^5 N/m^2 will be equivalent to

$$\frac{1.00\times10^5 \text{N}/\text{m}^2}{1.013\times10^3 \text{N}/\text{m}^2}\times760\text{mm Hg} = 750\text{mm Hg.}$$

Q. 2 : A gaseous mixture of 90 mole percent hydrogen and 10 mole percent deuterium at 25°C and a total pressure of 1 atm effuses through a small pin hole of area 0.30 mm^2. Calculate the composition of the initial gas that passes through.

Solution : The rate of molecular flow or effusion in molecules per-second is given by, dn/dt. Thus rate of flow of

$$H_2 = \frac{dn_{H_2}}{dt} \qquad (1)$$

where n_{H_2} is in $\frac{\text{molecules}}{\text{cm}^3} = n_{H_2}\bar{c}_{H_2}A$ and $\bar{c}$ is the average speed in cm/sec and A is the area in cm². The rate of effusion of

$$D_2 = \frac{dn_{D_2}}{dt} = n_{D_2}\bar{c}_{D_2}A \qquad (2)$$

The ratio of equation (1) to equation (2) is

$$\frac{dn_{H_2}/dt}{dn_{D_2}/dt} = \frac{n_{H_2}\bar{c}_{H_2}A}{n_{D_2}\bar{c}_{D_2}A}$$

$$= \frac{n_{H_2}}{n_{D_2}} \cdot \frac{\bar{c}_{H_2}}{\bar{c}_{D_2}}.$$

$\frac{n_{H_2}}{n_{D_2}} = \frac{\text{number of molecules of } H_2/\text{cm}^3}{\text{number of molecules of } D_2/\text{cm}^3}$. The volumes cancel in both the numerator and the denominator. To solve the problem molecules should be changed to moles.

$$\frac{\text{number of moles of } H_2}{\text{number of moles of } D_2} = \frac{\text{number of molecules of } H_2/\text{Avagadro\#}}{\text{number of molecules of } D_2/\text{Avagadro\#}}.$$

So the ratio of mole percent equals the ratio of molecular percent. Therefore,

$$\frac{dn_{H_2}/dt}{dn_{D_2}/dt} = \frac{90}{10}\frac{\bar{c}_{H_2}}{\bar{c}_{D_2}} \qquad (3)$$

But the rate of effusion is inversely proportional to the m½ in g/mole. The root-mean-square velocities are proportional to arithmetical mean velocities, and since the proportionality constant cancels out, equation (3) is written as

$$\frac{dn_{H_2}/dt}{dn_{D_2}/dt} = \frac{90}{10}\sqrt{\frac{M_{D_2}}{M_{H_2}}}$$

$$= \frac{90}{10}\sqrt{\frac{4.028}{2.016}}$$

$$= 9\sqrt{2.0}$$

$$= 12.7$$

Thus 12.7 molecules of H_2 per molecule of D_2 effuses, and the percent composition of the initial gas passing through will be

$$\text{mole \% } D_2 = \frac{1\times100}{12.7+1}$$
$$= 7.30\%.$$
$$\text{mole \% } H_2 = 100 - 7.30$$
$$= 92.7\%.$$

Q. 3 : Calculate the average velocity of Caesium atoms forming an atomic beam emerging from a small hole of a wall of an oven. The oven contains a sample of Caesium and is heated to 500°C.

Solution : Assume that the atoms impinging on the wall of the oven are doing so from the x-direction, so only their velocity components in that direction need be cosidered. The velocity of the atoms are thus between zero and infinity. The mean velocity $<V_x>$, of a particle is given by

$$< V_x > = \int_0^{\infty} V_x f(V_x) dV_x \qquad (1)$$

where $f(V_x)$ is the velocity distribution given as

$$f(V_x) = (m/2\pi kT)^{1/2} \exp(-\tfrac{1}{2}\, m\, V_x^2/kT) \qquad (2)$$

Also $\int_{-\infty}^{\infty} f(V_x) dV_x = 1$

The velocity component lies in the range $-\infty \le V_x \le +\infty$; hence, its probability must be unity.

In order to make the integral $\int_0^{\infty} f(V_x) dV_x = 1$, it is necessary to multiply the right-hand side of equation (2) by 2. Therefore from equation (1)

$$< V_x > = 2(m/2\pi kT)^{1/2} \int_0^{\infty} V_x \exp(-mV_x^2/2kT) dV_x$$
$$= 2(m/2\pi kT)^{1/2}(kT/m) = \sqrt{(2kT/\pi m)}.$$

For Caesium at 500° C,

$$< V_x > = \left[\frac{(2/\pi)\times(1.381\times10^{-23} JK^{-1})\times(773K)}{(132.9)\times(1.6605\times10^{-27} kg)}\right]^{1/2}$$

$$< V_x > = 175.5 m-s^{-1}.$$

Q. 4 : Air is estimated to be 20 mole percent o_2 and 80 mole percent N_2 and the total external surface area of an average human being is 2.0m². Estimate the number of collisions of air molecules per second with the surface of an average human being at 25°C and 1 atm.

Solution : Assume a one dimensional problem; consequently, only collisions in the positive x-direction will be considered. The mean velocity $\bar{c}_x$ is given by the integral of $c_x f(c_x)$ between the limits of $c_x = 0$ and $c_x = \infty$. That is,

$$\bar{c}_x = \int_0^\infty c_x f(c_x) dc_x \quad (1)$$

where $f(c_x)$ is some exponential function of c_x given by

$$f(c_x) = \left(\frac{m}{2\pi kT}\right)^{1/2} \exp\left(\frac{-\frac{1}{2} m c_x{}^2}{kT}\right) \quad (2)$$

Mutiply the right-hand side of (2) by 2 so that $\int_0^\infty f(c_x) dc_x = 1$

Equations (1) and (2) become

$$\bar{c}_x = 2(m/2\pi kT)^{1/2} \int_0^\infty c_x \exp\left(\frac{-mc_x^2}{2kT}\right) dc_x$$

$$= 2\left(\frac{m}{2\pi kT}\right)^{1/2} (kT/m)$$

$$= \sqrt{\frac{2kT}{\pi m}}$$

$$\bar{c}_x = \sqrt{\frac{2RT}{\pi M}}$$

The collision frequency, Z is given by

$$Z = \frac{1}{2} N\bar{c}$$

where N is the number of molecules and ½ is the fraction of molecules moving toward the surface.

$$Z = \frac{\frac{1}{2} \times 1 \text{ atm} \times 6.023 \times 10^{23} \text{ molecules mole}^{-1}}{82.06 \text{ cm}^3 \text{ atm deg}^{-1} \text{ mole}^{-1} \times 298 \text{deg}}$$

$$\times \sqrt{\frac{2 \times 8.314 \times 10^7 \times 298}{\pi}} \left(\frac{0.2}{\sqrt{32}} + \frac{0.8}{\sqrt{28}}\right)$$

$$= 2.88 \times 10^{23} \text{ collisions cm}^{-2} \text{ sec}^{-1}.$$

But the number of collisions = ZA, where A is the area.

$$ZA = (2.885 \times 10^{23} \text{collisions cm}^{-2} \text{ sec}^{-1})\left(\frac{100\text{cm}}{1\text{m}}\right)^2 (2.0\text{m}^2)$$

or

$$ZA = 5.77 \times 10^{27} \text{ collisions sec}^{-1}.$$

Q. 5. The concentration of hydrogen atoms in intergalactic space at a temperature of 7.8 × 10^{50}K is about 1 atom per 100 liter. Assuming the colision dimater of the hydrogen atom is 2.0Å caculate (a) the mean free path of an atom in light-years.

(b) The approximate average time, in years, between collisions of a given atom.

Solution : (a) The mean free path, λ, is a measure of the average distance that a molecule travels between successive collisions and is defined as

$$\lambda = \frac{\bar{c}}{Z}$$

where $\bar{c}$ is the mean speed of the atoms and Z is the collision frequency.

Equation (1) can also be written as

$$\lambda = \frac{\bar{c}}{2^{1/2}\pi\sigma^2\bar{c}c}$$

or

$$\lambda = \frac{1}{2^{1/2}\pi\sigma^2 c}$$

where c is the concentration

$$\lambda = \frac{1}{2^{1/2}\pi(10^{-5}\text{atom cm}^{-3})(4\times10^{-16}\text{cm}^2)}\left(\frac{1 \text{ light - yr}}{9.464\times10^{17}\text{cm}}\right)$$

$$\lambda = 59 \text{ light - yr}$$

(b) The mean speed is given by

$$\bar{c} = \sqrt{\frac{8RT}{\pi M}}$$

where M is the molecular weight of hydrogen and R is the gas constant. Inserting the respective values into the equation yields

$$\bar{c} = \sqrt{\frac{8(8.314)(10^7)(7.8\times10^5 \text{erg}(\text{g atom})^{-1}}{\pi(1.008\text{g}(\text{g atom})^{-1})}}$$

$$= \sqrt{1.64\times10^{14} \text{erg g}^{-1}}$$

$$\bar{c} = 1.28\times10^7 \text{cm sec}^{-1}$$

$$\text{Time, } t = \frac{\lambda}{\bar{c}}$$

$$= \frac{59 \text{light - yr}(9.464\times10^7 \text{cm}(\text{light - yr})^{-1})}{1.28\times10^7 \text{cm sec}^{-1}}\left(\frac{1 \text{ yr}}{3.156\times10^7 \text{ sec}}\right)$$

$$= 138.2\times10^3 \text{ yr}$$

or

$$t = 1.382 \times 10^5 \text{ yr.}$$

Q. 6. Calculate the number of moles of air in a column 1 cm² in area and rising from sea level to an altitude of 1.5×10^6 cm (about 10 miles), allowing for the barometric decrease in pressure. Assume air to be a uniform gas of molecular weight 29, and a uniform temperature of 0°C. Note that the atmospheric pressure is equal to the total weight of gas in a column of unit area and of infinite height. Conversion factors: R = 0.082 liter-atm/mole-deg = 8.31×10^7 ergs/more-deg = 1.98 cal/mole-deg ; 1 atm = 760 mm Hg = 1×10^6 dynes/cm².

Solution : The column of air at a uniform temperature is subject to a gravitational field of acceleration, g. Due to this field, the forces acting upon the lower and upper boundaries are not the same. The difference is equal to the weight of the fluid between the boundaries. Thus,

$$-df = \pi r^2 \rho g \, dh \qquad (1)$$

where f = force, ρ = density, dh = difference in height and g = gravity constant. But the difference in the force per unit area is equal to the pressure difference between the lower and upper boundaries. Therefore, equation (1) takes the form

$$\frac{-df}{\pi r^2} = -dP = \rho g dh \tag{2}$$

The minus sign indicates that the pressure decreases with increasing altitude. For ideal gases,

$$\rho = \frac{PM}{RT} \tag{3}$$

where M = molecular weight.

Using equation (3) appropriately, equation (2) becomes

$$-dP = \frac{PM}{RT} g dh$$

or

$$\frac{dP}{P} = -\frac{Mg}{RT} dh \tag{4}$$

Integrating equation (4) gives

$$\int_{P_1}^{P_2} \frac{dP}{P} = -\frac{Mg}{RT} \int_0^h dh$$

But P_1 = 1 atmosphere. Therefore

$$\int_1^{P_2} \frac{dP}{P} = -\frac{Mg}{RT} h$$

or

$$\ln P|_1^{P_2} = -\frac{Mg}{RT} h,$$

which becomes

$$\ln\left(\frac{P_2}{1}\right) = -\frac{Mg}{RT} h$$

$$= -\frac{(29)(980)(1.5 \times 10^6)}{(8.31 \times 10^7)(273)}$$

$$= -1.879$$

$$P_2 = e^{-1.879} = 0.153 \text{ atm}$$

Since the weight of a column of air extending down to sea level is 1×10^6 dynes cm^{-2}, the weight of a column extending down only to a given altitude will be 0.153×10^6 dynes cm^{-2}. Therefore, the weight of the column between the sea level and this given altitude is $1\times 10^6 - 0.153 \times 10^6 = 0.847 \times 10^6$ dynes cm^{-2}. The weight of this colum between the sea level and the given altitude is given by

$$w = nMg \quad (5)$$

where w = weight

n = number of moles.

Rearranging equation (5) to solve for n gives

$$n = \frac{w}{Mg} = \frac{(0.847\times10^6)}{(29)(980)}$$

$$n = 29.8 \text{ moles.}$$

Q. 7. Air is approximately 80% nitrogen and 20% oxygen (on a mole basis). If 4.0 g of hydrogen is added to a 22.4-liter flask maintained at 0°C and initially filled with air at 1 atm pressure, what will be the molecular weight (i.e., the average molecular weight) of the hydrogen air mxiture ?

Solution : There are 22.4 liters of air at STP (0°C and 1 atm). Therefore, there is 1 mole of air. Also, there are 4 gm of hydrogen, which is equivalent to

$$\frac{4 \text{ gm}}{2 \text{ gm / mole}} = 2 \text{ moles.}$$

By definition,

$$M_{average} = \left[\sum_{i=1}^{3} n_i M_i\right] \Big/ \left[\sum_{i=1}^{3} n_i\right]$$

wheere n_1 = moles of oxygen in air

n_2 = moles of nitrogen in air

n_3 = moles of hydrogen

M_1 = molecular weight of oxygen

M_2 = molecular weight of nitrogen

M_3 = molecular weight of hydrogen.

$$M_{average} = \frac{n_1M_1 + n_2M_2 + n_3M_3}{n_1 + n_2 + n_3}$$

$$= \frac{(.2)(32) + (.8)(28) + 2(2)}{.2 + .8 + 2} = \frac{32.8}{3} \text{gm / mole}$$

$$= 10.9 \text{ gm / mole.}$$

Q. 8. Hydrogen gas will dissociate into atoms at a high enough temperature, i.e., $H_2 = 2H$. Assuming ideal gas behavior for H_2 and H, what shoule be the density of hydrogen at 2000°C if it is 33% dissociated into atoms ? The pressure is 1 atm.

Solution : Use as a basis 1 mol hydrogen before dissociation (which has a weight of 2 gms). After 0.33 moles have dissociated, then we have

0.67 mole H_2 left.

Therefore, there are 0.66 moles of H. The total number of moles are 0.67 + 0.66 = 1.33 moles.

For a mixture of gases, the density

$$\rho = \frac{PM_{average}}{RT} \quad \text{where} \quad M_{average} = \frac{\text{total weight}}{\text{total moles}} = \frac{2\ \text{gm}}{1.33\ \text{mol}}$$

Note that the units of $M_{average}$ are consistent with the units of molecular weight. This is very important.

Therefore,

$$\rho = \frac{(1)(2/1.33)}{(0.0082)(2273)} = 8.07 \times 10^{-3}\ \text{gm/liter}\ .$$

Q. 9. Calculate the minimum work that must be expended in the compression of 1 kg of ethylene from $10^{-1}m^3$ to $10^{-2}m^3$ at a constant temperature of 300°K, assuming the gas to be

(a) ideal (b) Van der Waals.

Solution : (a) The work done in an isothermal compression of an ideal gas is given by

$$W = -\int PdV \qquad (1)$$

But $P = \frac{nRT}{V}$. Substituting P = nRT/V into equation (1) gives

$$W = -\int nRT\frac{dV}{V}$$

Since nRT is constant,

$$W = -nRT\int_{V_1}^{V_2} \frac{dV}{V}$$

or

$$W = -nRT\ \ln\frac{V_2}{V_1} \qquad (2)$$

The number of moles of C_2H_4 are

$$n = (1\text{kg } C_2H_4)\left(\frac{1000\text{g}}{1\text{kg}}\right)\left(\frac{1\text{mole } C_2H_4}{28.05\text{g}}\right)$$

$$n = 35.65 \text{ mole.}$$

Hence equation (2) becomes

$$W = -(35.65 \text{ moles})(8.314\ \text{J}^\circ\text{K}^{-1}\text{mol}^{-1})(300^\circ\text{K})\ln\frac{10^{-2}\text{ m}^3}{10^{-1}\text{ m}^3}$$

$$= 2.047\times10^5 \text{ J}$$

or

$$W = 204.7 \text{ kJ.}$$

(b) Again the work done is given by

$$W = -\int PdV.$$

For a gas that obeys the Van der Waals equation, P is given by

$$P = \frac{nRT}{V-nb} - \frac{n^2a}{V^2}$$

where a and b are Van der Waals constants given as 4.47 ℓ^2 atm mol^{-2} and 0.0571 ℓ mol^{-1} respectively.

Substituting for P gives

$$W = -\int PdV = -\int\left[\frac{nRT}{V-nb} - \frac{n^2a}{V^2}\right]dV$$

or

$$W = -\int_{V_1}^{V_2}\frac{nRT}{V-nb}dV + \int_{V_1}^{V_2}\frac{n^2a}{V^2}dV$$

Multiplying equation (5) through by $V^2(V-nb)^2$ gives

$$nRTV^2(V-nb) - nRTV^3 + n^2a(V-nb)^2 = 0$$

or

$$nRTV^2(V-nb) - nRTV^3 = -n^2a(V-nb)^2.$$

Therefore,

$$RT = \frac{a}{b}\left[\frac{V-nb}{V}\right]^2 = \frac{a}{b}\left[\frac{V}{V} - \frac{nb}{V}\right]^2$$

$$=\frac{a}{b}\left[1-\frac{nb}{V}\right]^2 \tag{6}$$

At the Boyle point $T = T_B$, $P = 0$ and $V = \infty$, so $RT_B = a/b$

or

$$T_B = \frac{a}{Rb}.$$

Another method of doing this is to set $n/V = c$. So, the van der Waals equation of state becomes

$$P = \frac{cRT}{1-bc} - ac^2 \tag{7}$$

From equation (7),

$$PV = n\left[\frac{RT}{1-bc} - ac\right] \tag{8}$$

Using the chain rule to differentiate equation (8) partially with respect to P gives

$$\left(\frac{\partial(PV)}{\partial P}\right)_T = \left(\frac{\partial(PV)}{\partial c}\right)_T\left(\frac{\partial c}{\partial P}\right)_T = n\left[\frac{bRT}{(1-bc)^2} - a\right]\left(\frac{\partial c}{\partial P}\right)_T. \tag{9}$$

Observe that $\left(\frac{\partial c}{\partial P}\right)_T = \frac{1}{(\partial P/\partial c)_T} = \left(\frac{\partial P}{\partial c}\right)_T^{-1}$. From equation (7),

$$\left(\frac{\partial P}{\partial c}\right)_T = \left[\frac{bcRT}{(1-bc)^2} + \frac{RT}{1-bc} - 2ac\right]^{-1} \tag{10}$$

Observe that equation (10) reduced to

$$RT \text{ as } P \to 0 \text{ and } c \to 0 \text{ or } \left(\frac{\partial c}{\partial P}\right) = 1/RT.$$

Since $1/RT \neq 0$ and $n \neq 0$, the term in the squre brackets in equation (9) must vanish at $c = 0$ and $T = T_B$. That is,

$$bRT_B - a = 0 \tag{11}$$

Solving for T_B changes equation (11) to

$$T_B = \frac{a}{bR}.$$

Q. 10. Gases A, B, C, and D obey the van der Waals equation, with a and b values as given (liter-atm system of units):

	A	B	C	D
a	6	6	20	0.05
b	0.025	0.15	0.10	0.02

Which gas has the highest critical temperature, the largest molecules, the most nearly ideal behavior at STP ?

Solution : Looking at $\frac{P_cV_c}{RT_c} = \frac{(a/27b^2)(3b)}{(R)(8a/27Rb)}$ it is seen that T_c is proportional to a/b. If this ratio is computed for the various gases, than gas A, a/b = 6/0.025 = 240; For gas B, a/b = 6/0.15 = 40. For gas C, a/b = 20/0.10 = 200 and for gas D, a/b = 0.05/0.02 = 2.5. These values show that gas A (with the highest a/b ratio) has the highest T_c.

The constant b gives a measure of molecular volume (that is, V_c = 3b). Therefore, gas B has the largest molecules. At STP, the gas that has the lowest critical temperature and pressure will be most ideal. Gas D has the lowest a/b ratio, hence lowest T_c. The same gas also gives the lowest a/b^2 ratio, hence the lowest critical pressures P_c.

Q. 11. The van der Waals constants for gases A, B, and C are :

Gas	a, liters²-atm/mole	b, liters/mole
A	4.0	0.027
B	12.0	0.030
C	6.0	0.032

Which gas has (a) the highest critical temperature, (b) the largest molecular volume, and (c) most ideal general behavior around STP ?

Solution : The van der Waals equation is

$$\left(P + \frac{n^2a}{V^2}\right)(V - nb) = nRT$$

where P = pressures
T = temperature
V = volume
n = number of moles.

a and b are van der Waals constants.

(a) It is knwon that T_c, the critical temperature is proportional to the ratio a/b. Computing the a/b ratios for the three gases yields :

for gas A $\dfrac{4.0}{0.027} = 148.15$

for gas B $\dfrac{12.0}{0.030} = 400$

for gas C $\dfrac{6.0}{0.032} = 187.5$

Since the ratio for gas B is the largest, gas B has the largest T_c. This question could have been answered by mere inspection. It can be seen that the constant b has almost the same value for the three gases. So, the larger a value means greater intermolecular forces. A larger T_c would be needed to overcome them. Since gas B has the largest a value, its T_c will be the largest.

(b) The constant b is proportional to V_c. Therefore gas C has the largest V_c.

(c) The b values are almost the same. Therefore ideality will be determined by the a values. Looking at the original equation, it can be seen that it is near ideality for values of the constants close to zero. Since gas A has the smallest a value, it should be the most ideal in behavior. Remember that for ideality, the expression is PV = nRT.

Q. 12. Given the following data for a certain nonideal gas at 25°C :

ρ/P, g/liter-atm :	10	11	10	(r = density)
P, atm :	1	10	20	

The critical pressure of this gas must then be (greater than 10 atm, greater than 20 atm, between 1 and 20 atm, between 1 and 10 atm, less than 20 atm, can't tell).

Solution : Density $= \dfrac{\text{weight}}{\text{volume}} = \dfrac{W}{V}$

$$\frac{\text{density}}{\text{pressure}} = \frac{W/V}{P} = \frac{\rho}{P} \tag{1}$$

where $\rho = W/V$.

$$\text{Compressibility factor} = \frac{PV}{RT} \tag{2}$$

Therefore, comparing (1) and (2), it is observed that ρ/P is inversely proportional to the compressibility factor at a given temperature, so the maximum in ρ/P means a minimum in PV/RT. In plots of PV/RT vs. P/P_c, such a minimum occurs only for $T < T_c$ and $P > P_c$ where T_c and P_c are critical temperature and pressure, respectively. The data tells us that this minimum is between 1 atm and 20 atm. Therefore P_c could be greater or less than 1 atm and greater or less that 10 atm, but it must be less than 20 atm.

Q. 13 : The critical temperature and pressure for NO gas are 177°K and 64 atm, respectively, and for CCl_4 they are 550°K and 45atm, respectively. Which gas (a) has the smaller value of the van der Waals constant b, (b) the smaller value of the van der Waals constant a, (c) is most nearly ideal in behavior at 300°K and 10 atm ?

Solution : For any gas that obeys van der Waals equation, $V_c = 3b$, $P_c = a/27b^2$ and $T_c = 8a/27bR$.

(a) If we take the ratio T_c/P_c, we have

$$\left[\frac{8a}{27bR}\right] \times \left[\frac{27b^2}{a}\right] = \frac{8b}{R}$$

Therefore b is directly propoortional to T_c/P_c. Also it is observed that $V_c = 3b$ is proportional to T_c/P_c.

T_c/P_c for NO gas = 177/64 = 2.766.

T_c/P_c for CCl_4 = 550/45 = 12.22.

Therefore NO should have the smaller b value.

(b) The constant a is proportional to $P_cV_c^2$ or to T_c^2/P_c. Therefore it will be smaller for NO.

(c) No should be more ideal sine 300°K is above its critical temperature, but below that of CCl_4.

Q. 14. A nonideal gas of molecular weight 150 obeys the van der Waals equation ; its critical pressure and temperature are 100 atm and 100°C, respectively.

(a) The compressibility factor PV/RT will be greater than unity at (500 atm and 80°C, 500 atm and 120°C, 50 atm and 60°C, 50 atm and 120°C, none of these).

(b) Calculate the value of the compressibility factor at the critical point.

Solution : The van der Waals equation is

$$\left(P+\frac{n^2a}{V^2}\right)(V-nb)=nRT \qquad (1)$$

(a) For the compressibility factor to be greater than unity, what is needed is a high pressure and temperature. By this it is meant that the required temperature and pressure must both be greater than the values at the critical point. Hence 120°C and 500 atm is the correct answer.

(b) Compressibility factor at the critical point $= \frac{P_cV_c}{RT_c}$, where the subscript c means the properties at the critical point. For a gas that obeys the van der Waals equation,

$$T_c = \frac{8a}{27Rb},$$

$V_c = 3b$ and $P_c = \frac{a}{27b^2}$ where a and b are constants (as in equation (1)).

Now after substituting these values into $\frac{P_cV_c}{RT_c}$, it yields

$$\frac{\left(\frac{a}{27b^2}\right)(3b)}{(R)\left(\frac{8a}{27Rb}\right)}$$

$$\text{Compressibility factor} = \frac{3ab}{27b^2}\times\frac{27Rb}{8aR}=\frac{3}{8}.$$

2

First Law of Thermodynamics and Thermochemistry

Q. 1. For the system described by the following diagram, is the forward reaction exothermic or endothermic ?

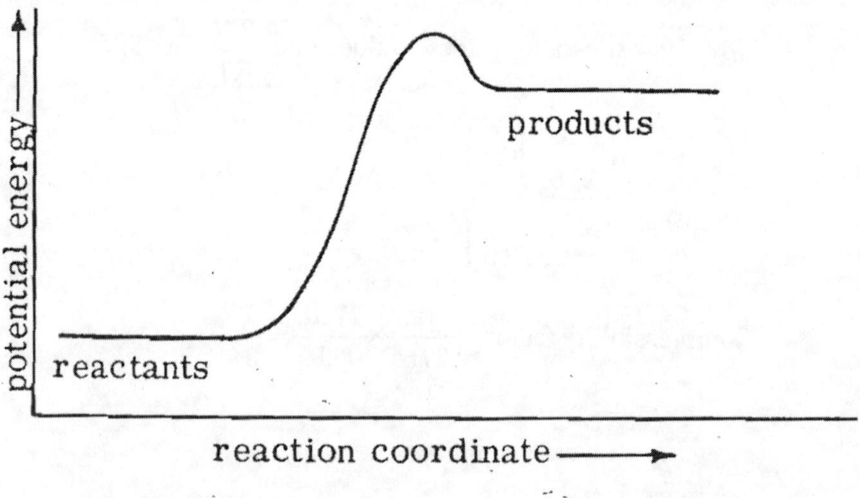

Solution : By observing the diagram, we see that the products are at a higher potential energy than the reactants, thus energy must have been absorbed during the course of the reaction. The only mechanism by which this can occur is absorption of heat and subsequent conversion of heat into potential energy. Hence, the reaction absorbs heat and is therefore endothermic.

Q. 2. A steam engine is operated reversibly between two heat reservoirs, a parimary reservoir and a secondary reservoir at 35°C. Determine the efficiency of the engine.

Solution : By definition, the primary reservoir is the reservoir from which the engine absorbs heat and the secondary reservoir is the reservoir to which the engine releases heat. Since the engine is steam operated, the primary reservoir must be a source of steam (such as a boiler) operating at a temperature $T_p = 100°C$ ($T_p = 100 + 273 = 373°K$). The secondary reservoir operates at a temperature $T_s = 35 + 273 = 308°K$. The efficiency of such an engine is defined as

$$\text{efficiency} = \frac{T_p - T_s}{T_p}$$

where T_p and T_s are in degrees Kelvin. Thus, the efficiency of the engine is

$$\text{efficiency} = \frac{T_p - T_s}{T_p} = \frac{373°K - 308°K}{373°K} = 0.174 \text{ or } 17.4\%.$$

Q. 3. Given for hydrogen peroxide, H_2O_2 :

$HO + OH(g) \rightarrow 2OH\ (g)\ \Delta H°_{diss} = 51$ Kcal/mole

From this value and the following data calculate :

(a) $\Delta H°_f$ of OH (g). (b) C–O bond energy; $\Delta H°_{diss}$ in CH_3OH (g). $\Delta H°_f$ of H_2O_2 (g) = –32.58 Kcal/mole. $\Delta H°_f$ of CH_3 (g) = 34.0 Kcal/mole. $\Delta H°_f$ of CH_3OH (g) = –47.96 Kcal/mole.

Solution : (a) $\Delta H°_{diss}$ for this reaction is equal to the $DH°_f$ of reactants subtracted from the $\Delta H°_f$ of the products, where $\Delta H°$ equals enthalpy or heat content.

$\Delta H°_{diss} = 2 \times \Delta H°_f OH - \Delta H°_f HO - OH = 51$ Kcal/mole

$2 \times \Delta H°_f OH = 51$ Kcal/mole $+ \Delta H°_f$ HO–OH

$$\Delta H_f^\circ OH = \frac{51 \text{ Kcal/mole} - 32.58 \text{ Kcal/mole}}{2}$$

$$\Delta H_f^\circ OH = \frac{18.42 \text{ Kcal/mole}}{2} = 9.21 \text{ Kcal/mole.}$$

(b) CH_3OH dissociates by breaking the C–O bond :

$CH_3OH\ (g) \rightarrow CH^+_3\ (g) + OH^-\ (g)$

Therefore, the bond energy of the C–O bond equals $\Delta H°_{diss}$.

$\Delta H°_{diss}$ equals the sum of the ΔH's of the products minus the ΔH's of the reactants. Thus,

bond energy of C–O= $\Delta H°_f$ of CH^+_3 (g) + $\Delta H°_f$ of OH– (g) – $\Delta H°_f$ of CH_3OH (g).

bond energy of C–O = 34.0 Kcal/mole + 9.21 Kcal/mole + 47.96 Kcal/mole
= 91.17 Kcal/mole.

Q. 4. What is the heat capacity of uranium ? Its atomic heat is 26 J/mole and its atomic weight is 238 g/mole.

Solution : The atomic heat of a solid element at room temperature is defined as the amount of heat required to raise the temperature of one mole of an element by one degree Celsius. The heat capacity, or specific heat, is the amount of heat required to raise the temperature of one gram of a substance by one degree Celsius. These two quantities can be related by the equation,

atomic heat = atomic weight × heat capacity.

Thus, to find the heat capacity of uranium, substitute the given values of atomic heat and weight.

Heat capacity of uranium = 26 J/mole C°/238 g/mole
= 0.11 J/g°C.

Q. 5. Determine ΔH° for the following reaction of burning ethyl alcohol in oxygen :

$$C_2H_5OH\ (\ell) + 3O_2\ (g) \rightarrow 2CO_2\ (g) + 3H_2O\ (\ell)$$

$\Delta H°_f$ of C_2H_2OH (ℓ) = – 65.9 Kcal/mole

$\Delta H°_f$ of CO_2 (g) = –94.1 Kcal/mole

$\Delta H°_f$ of H_2O (ℓ) = –68.3 Kcal/mole

Solution : The heat reaction (ΔH°) may be found from the heats of formation ($\Delta H°_f$) by subtracting the sum of the heats of formation of all reactants from the sum of the heats of formation of all products. The heat of formation of all pure elments is zero. When more than one mole of a compound is either reacted or formed, the heat of formation is multiplied by the stoichiometric coefficient for the specific compound in the equation.

heat of reaction = (2 × $\Delta H°_f$ of CO_2 + 3 × $\Delta H°_f$ of H_2O) – ($\Delta H°_f$ of C_2H_2OH + 3 moles × $\Delta H°_f$ of O_2)

heat of reaction = [2 moles × (–94.1 Kcal/mole) + 3 moles ×

$$(-68.3 \text{ Kcal/mole}) - 1 \text{ mole} \times (-65.9 \text{ Kcal/mole} + 3 \text{ moles} \times 0)$$
$$= -327.2 \text{ Kcal.}$$

Q. 6. Find the $\Delta H°$, heat of reaction, for the hydrolysis of urea to CO_2 and NH_3. Although the reaction does not proceed at a detectable rate in the absence of the enzyme, urease, ΔH may stil be computed from the following thermodynamic data :

Compound	$\Delta H°_f$ (Kcal/mole)
Urea (aq)	–76.30
CO_2 (aq)	–98.69
H_2O (ℓ)	–68.32
NH_3 (aq)	–19.32

The hydrolysis reaction is

$$H_2O(\ell) + \underset{\text{(urea)}}{N_2N - CO - NH_2} \rightarrow CO_2(aq) + 2NH_3(aq)$$

Solution : The heat of reaction, $\Delta H°$, may be found from the heats of formation, $\Delta H°_f$, of the reactants and products.

Heat of formation may be defined as the heat absorbed or evolved in the synthesis of one mole of a compound from its elements, all components being in their standard states.

The heat of reaction is the sum of the heats of formation of the products minus the sum of the heats of formation of the reactants. In other words,

$$\Delta H_{reaction} = \Sigma \Delta H_{formation\ of\ products} - \Sigma \Delta H_{formation\ of\ reactants}$$

Therefore, for the hydrolysis of urea to CO_2 and NH_3

$$\Delta H_{reaction} = \left((\Delta H°_{f,CO_2} + 2\Delta H°_{f,NH_3})\right) - \left(\Delta H°_{f,H_2N-CO-NH_2} + \Delta H°_{f,H_2O}\right)$$

Given these values, one can substitute to obtain:

$$\Delta H_{reaction} = [(-98.69 - 2 \times 19.32) - (-76.30 - 68.32)]$$

From this equation one obtains

$$\Delta H_{reaction} = 7.29 \text{ Kcal/mole.}$$

Q. 7. It is known that 1 ton of TNT can produce 4.6×10^9 J of energy. A chemist wanted to obtain this much energy from acetylene (C_2H_2) and oxygen (O_2). How much C_2H_2 and O_2 would be required ? The $\Delta H° = -2510$ KJ for the reaction between C_2H_2 and O_2.

Solution : The equation for this reaction is

$2C_2H_2 + 5O_2 \rightarrow 4CO_2 + 2H_2O$, $\Delta H° = -2510$ KJ. In a chemical reaction, energy is usually liberated or absorbed. The $\Delta H°$ enthalpy, expresses this energy liberation or absorption quantitatively. The negative sign denoted the release of energy.

Two moles of acetylene and 5 moles of oxygen react to form 2510 KJ or 2510×10^3 J of energy. Similarly, 2x moles of acetylene and 5x moles oxygen produce 2.510×10^6 xJ of energy. In this case, the energy released must equal 4.6×10^9 J. Therefore,

$$2.510 \times 10^6 x = 4.6 \times 10^9, \text{ or } x = 1.83 \times 10^3.$$

From the equation for the reaction one knows that 2 moles of C_2H_2 and 5 moles of O_2 react to produce 2510 KJ. The calculations already done show that to produce as much energy as 1 ton of TNT there must be $1.83 \times 10^3 \times 2$ moles of C_2H_2 and $1.83 \times 10^3 \times 5$ moles of O_2 to form 4.6×10^9 J. To solve for the amount of each needed, multiply the number of moles by the molecular weight of the substance. (MW of $C_2H_2 = 26$).

$$\text{wt of } C_2H_2 = (1.83 \times 10^3 \times 2) \text{ moles} \times 26 \text{ g/moles}$$
$$= 9.52 \times 10^4 \text{ g} = 9.52 \times 10^4 \text{ g} \times 1.10 \times 10^{-6} \text{ tons/g}$$
$$= .1044 \text{ tons.}$$

(MW of $O_2 = 32$)

$$\text{wt of } O_2 = (1.83 \times 10^3 \times 5) \text{ moles} \times 32 \text{ g/mole}$$
$$= 5.86 \times 10^4 \text{ g} = 5.86 \times 10^4 \text{ g} \times 1.10 \times 10^{-6} \text{ tons/g}$$
$$= 0.646 \text{ tons.}$$

Q. 8. Calculate ΔH for the reaction at 25°C.

$CO\ (g) + \frac{1}{2}O_2\ (g) \rightarrow CO_2\ (g)$ $\Delta E = -67.4$ Kcal.

Solution : The change in enthalpy (ΔH) is defined by the first law of thermodynamics as the sum of the change of the internal energy (ΔE) and the work done by a particular system. Work is defind as the product of the pressure (P) and the change in volume (ΔV) of a system, where only pressure-volume work is done. Much of the experimentation done by chemists is pressure-volume work because the work is usually done in an open container in the laboratory. The pressure, in this case, is constant and the volume of the materials is allowed to change.

$$\Delta H = \Delta E + P\Delta V$$

In this problem, one is given ΔE but not P or ΔV, so that one must calculate PDV in some other way. The Ideal gas Law is defined as:

$$P\Delta V = \Delta nRT,$$

where P is the pressure, ΔV is the change in volume, Δn is the change in the number of moles reacting, R is the gas constant (1.99 cal/mole–°K), and T is The absolute temperature. Here, one knows R and can calculate T and Δn. Therefore, ΔnRT can be substituted for $P\Delta V$ in the equation to find ΔH. Δn is found by subtracting the number of moles reacting from the number of moles formed as products. Here, 1 mole of CO and 0.5 mole of O_2 form 1 mole of CO_2.

$\Delta n = 1 \text{ mole } CO_2 - (1 \text{ mole } CO + 0.5 \text{ mole } O_2)$
$= -0.5$ mole.

The absolute temperature is found by adding 273 to the temperature in °C.

$$T = 25 + 273 = 298°K.$$

Therefore, $\Delta nRT = (-0.5 \text{ mole}) \times 1.99 \text{ cal/mole–°K} \times 298°K$
$= -0.3$ Kcal.

One is now ready to find ΔH, via substitution.

$\Delta H = \Delta E + \Delta nRT$
$\Delta E = -67.4$ Kcal
$\Delta nRT = -0.3$ Kcal
$\Delta H = -67.4 \text{ Kcal} + (-0.3 \text{ Kcal}) = -67.7$ Kcal.

Q. 9. For the following exothermic reaction at 25°C,

$$C_6H_6\ (\ell) + 7\tfrac{1}{2}\ (g) \rightarrow 6CO_2\ (g) + 3H_2O\ (\ell),$$

the change in energy, DE, is –780 Kcal/mole. Find the change in enthalpy, ΔH.

Solution : The solution is based on the following equation for the enthalpy change

$$\Delta H = \Delta E + \Delta nRT,$$

where Δn is the number of moles of gaseous products minus the number of moles of gaseous reactants, R is the gas constant (1.987 cal/deg–mole), and T is the absolute temperature. Since there are 6 moles of gaseous products ($6CO_2$ (g)) and 7½ moles of gaseous reactants ($7\frac{1}{2}O_2$ (g)), $\Delta n = 6 - 7\frac{1}{2} = -1.5$. The absolute temperature is T = 25° + 273 = 298°K.

Hence, $\Delta H = \Delta E + \Delta nRT$
$= -780 \text{ Kcal} + (-1.5 \text{ mole})(1.987 \text{ cal/deg–mole})$
$\times (298°K)$

= –780 Kcal – 888 cal

= – 780 Kcal – 0.888 Kcal

≅ –781 Kcal.

Q. 10. Calculate $C_p - C_v$ for water at 25°C given that the coefficient of thermal expansion of water is $257.05 \times 10^{-6} K^{-1}$ and the compressibility is 45.24×10^{-6} bar^{-1}. The density of water is $0.997075 \times 10^3 Kg\ m^{-3}$.

Solution : The difference between the heat capcities, $C_P - C_V$ is given by

$$C_P - C_V = \frac{\alpha^2 VT}{\beta} \quad (1)$$

where α = coefficient of thermal exapansion

β = compressibility factor

V = volume

and T = temperature

1 bar = 0.987 atm

Substituting the respective values into their appropriate places, equation (1) becomes

$$C_P - C_V = \left(\frac{(257.05\times10^{-6}K^{-1})^2(18.015\times10^{-3}kg\ mol^{-1}}{(45.24\times10^{6}\ bar^{-1})(0.987 atm\ bar^{-1})^{-1}}\right) \times\left((0.997075\times10^{3}\ kg\ m^{-3})^{-1}(298K)\right)$$

$= 7.76 \times 10^{-6} m^3$ atm $mol^{-1}K^{-1}$

$= 0.786$ J $mole^{-1}K^{-1}$

Q. 11. Consider the following reaction at 25°C

$$C(s) + 2H_2O\ (g) = CO_2\ (g) + 2H_2\ (g)$$

Calculate the heat capacity at constant pressure, C_P, for each of the substances in the reaction. For $H_2O(g)$, assume the contribution to C_P from the vibrational mode is 20%.

Solution : For C(s), assume Dulong and Petit's law. Therefore, the value of C_P is approximately 6 cal $K^{-1}mol^{-1}$. For H_2O (g), assuming 20% of vibrational contribution to C_P,

$$C_P = R + \left(\frac{3R}{2}\right)_{(trans)} + \left(\frac{3R}{2}\right)_{(rot)} + 0.2(3R)_{(vib)}$$

$$= 4.6R$$

$$= (4.6)(1.987 \text{ cal mol}^{-1} {}^{\circ}\text{K}^{-1})$$

$$= 9.1 \text{ cal}^{\circ}\text{K}^{-1}\text{mol}^{-1}$$

For CO_2 (g) : This is a linear molecule.

$$\therefore C_P = R + \left(\frac{3R}{2}\right)_{(trans)} + (2R/2)_{(rot)} + 0.2(4R)(v/b)$$

$$= 4.3R$$

$$= 8.5 \text{ cal } {}^{\circ}\text{K}^{-1} \text{ mol}^{-1}$$

for H_2 (g) :

$$C_P = R + (3R/2)_{(trans)} + (2R/2)_{(rot)} + 0.2\ (R)_{(vib)}$$

$$= 3.7R$$

$$= 7.3 \text{ cal } {}^{\circ}\text{K}^{-1}\text{mol}^{-1}$$

Q. 12. Given the reaction

$$CO\ (g) + H_2O\ (g) = CO_2\ (g) + H_2(g),\ \ \Delta H_{298^{\circ}K} = -10 \text{ kcal}$$

Calculate ΔH at 1000°K.

The heat capacities are

$C_P\ (CO) = 6.60 + 1.0 \times 10^{-3}T$

$C_P\ (H_2) = 6.6 + 1.0 \times 10^{-3}T$

$C_P\ (H_2O) = 7.3 + 2.0 \times 10^{-3}T$

$C_P\ (CO_2) = 7.3 + 3.0 \times 10^{-3}T$

Solution : The equation

$$\Delta H_{T_2} = \Delta H_{298^{\circ}K} + \int_{T_1}^{T_2} \Delta C_P dT \qquad (1)$$

is to be used

where ΔH = Enthalpy change

ΔC_P = Heat Capacity change

$$\Delta C_P = \underset{\text{producs}}{\Sigma C_P} - \underset{\text{reactants}}{\Sigma C_P}$$

That is,

$$\Delta C_p = \Sigma(CP(CO_2) + C_p(H_2) - C_p(CO) - C_p(H_2O))$$

Therefore,

$$\Delta CP = 7.3 + 3.0 \times 10^{-3}T$$
$$+ 6.6 + 1.0 \times 10^{-3}T$$
$$-6.60 + 1.0 \times 10^{-3}T$$
$$-7.3 + 2.0 \times 10^{-3}T$$

$$\Delta C_p = 0 + 10^{-3}T$$

From equation (1),

$$\Delta H_{1000°K} = \Delta H_{298°K} + \int_{298}^{1000} (10^{-3}T)dT$$

$$= \Delta H_{298°K} + \frac{10^{-3}}{2}(T_2^2 - T_1^2)$$

$$= -10\text{kcal} + \frac{10^{-3}}{2}(1000^2 - 298^2)$$

$$= -10\text{kcal} + 0.46$$

$$\therefore \quad \Delta H_{1000°K} = -9.54 \text{ kcal.}$$

Q. 13. Predict the heat of fusion for LiCl at 883°K given the following :

$$Li(liq) + \frac{1}{2}Cl_2(g) = LiCl(liq) \qquad \Delta H°_{883} = -92.347\frac{\text{koal}}{\text{mole}}$$

$$Li(liq) + \frac{1}{2}Cl_2(g) = LiCl(s) \qquad \Delta H°_{883} = -97.105\frac{\text{kcal}}{\text{mole}}$$

Solution : Hess's law states that if a series of reactions having known ΔH values can be arranged in such a way that when added up, they produce a desired reaction, the ΔH values for the desired reaction will be equal to the sum of the ΔH values for the individual reactions. The formation of LiCl (liq) from LiCl (s) illustrates the law. The heat of formation is the enthalpy change for a chemical reaction in which a compound is produced from its elements, both products and reactants being in their natrual states.

By definition,

$$\Delta H_T^\circ(\text{reaction}) = \sum_i^{\text{Products}} n_i \Delta H_T^\circ(\text{formation } i) - \sum_j^{\text{reactants}} n_j \Delta H_T^\circ(\text{formation } j)$$

where n_i and n_j are the number of moles of products and reactants respectively

$\therefore$ For the reaction

$LiCl(s) = LiCl\,(liq)$

and using the equation above,

$$\Delta H^{\circ}_{883} = (1)(-92.374) - (1)(-97.105)$$

$$= (4.731\text{ kcal})\left(\frac{4.184 \times 10^3\text{ J}}{1\text{kcal}}\right)\left(\frac{1\text{kJ}}{1000\text{J}}\right)$$

$$\Delta H^{\circ}_{883} = 19.795\text{kJ}.$$

Q. 14. Calculate the internal energy change for the reaction

$$\frac{1}{2}N_2(g) + \frac{1}{2}O_2(g) + \frac{1}{2}Cl_2(g) = NOCl(g)$$

at 25°C given that the gases are ideal. The enthalpy ΔH, for the formation of NOCl (g) is 12.57 kcal mol^{-1} at 25°C.

Solution : From the first law

$$\Delta E = \Delta H - \Delta(PV)$$

The work done against the atmosphere is $(n_b - n_a)$ RT:

where n_{b_g} = number of moles of gaseous product

n_{a_g} = number of moles of gaseous reactants

$$\therefore\ \Delta E = \Delta H - (n_{b_g} - n_{a_g})RT \qquad (1)$$

$$(n_{b_g} - n_{a_g}) = \Delta n_g = n_{NOCl} - \frac{1}{2}n_{N2} - \frac{1}{2}n_{O_2} - \frac{1}{2}n_{Cl_2}$$

$$= 1 - \frac{1}{2} - \frac{1}{2} - \frac{1}{2}$$

$$= -\tfrac{1}{2}\text{ mole}$$

Substituting this into equation (1) gives

$$\Delta E = (12.57\text{kcal})\left(\frac{4.184\text{kJ}}{1\text{kcal}}\right) - \left(-\frac{1}{2}\text{mole}\right)\left(\frac{8.314 \times 10^{-3}\text{kJ}}{1\cdot\text{mole}\ {}^{\circ}\text{K}}\right)(298^{\circ}\text{ K})$$

$$= (52.59 + 1.24)\text{kJ}$$

or

$$\Delta E = 53.83\text{ kJ}$$

Q. 15. What is ΔE, q, w and ΔH for the reversible compression of two moles of an ideal gas at 25°C from 1.00 atm to 100.0 atm ?

Solution : The internal energy of an ideal gas does not change with the distance between molecules becuase there is no interaction between them. At constant temperature the internal energy is independent of volume also. Since the nergy is not changed, the enthalpy will not change.

$$\therefore \quad \Delta E = \Delta H = 0$$

and $$q = w = \int_{V_1}^{V_2} PdV.$$

But $$P = \frac{nRT}{V}$$

Therefore,

$$q = w = \int_{V_1}^{V_2} PdV = \int_{V_1}^{V_2} \frac{nRT}{V} dV.$$

$$q = nRT \int_{V_1}^{V_2} \frac{dV}{V}.$$

The nRT term is outside the integral because it is a constant (independent of the volume).

Solving the integral,

$$q = nRT \ln V \Big|_{V_1}^{V_2}$$

$$= nRT(\ln V_2 - \ln V_1)$$

But $\ln a - \ln b = \ln (a/b)$

$$q = nRT(\ln V_2 - \ln V_1) = nRT \ln \frac{V_2}{V_1}$$

At constant temperature, Boyle's law states that

$$P_1 V_1 = P_2 V_2$$

or $$\frac{V_2}{V_1} = \frac{P_1}{P_2}$$

$$\therefore \quad q = w = nRT \ln \frac{P_1}{P_2}$$

$$= (2.00 mol)(8.314 J\ mol^{-1} K^{-1})(298° K) \ln \frac{1.00}{100.0}$$

$$= -22800 J = -22.8 kJ$$

3

Second Law of Themodynamics

Q. 1. A heat pump with a coefficient of performance given by

$$\text{COP} = -\frac{q_1}{w} = \frac{T_1}{T_2 - T_1} = 7.0$$

is used to pump an amount of heat q_1 at a temperature T_1, from the cold outdoors to the interior of a house at a temperature T_2. Calculate q_2/W, the ratio of the heat liberated at the hihger temperature T_2 to the work done.

Solution : A heat pump is essentialy a refrigerator operated in reverse; that is, it is used to heat a system at a higher temperature by puming heat from the surroundings at a lower temperature. Thus a heat pump could be used to warm a house by pumping heat from the cold outdoors and delivering it to the warm interior of the house. The advantage of this method of heating is that the maximum efficiency of a refrigerator is much greater than that of other methods of heating.

$$\text{COP} = -\frac{q_1}{w} = \frac{T_1}{T_2 - T_1} \qquad (1)$$

where q = quantity of heat

w = work done.

Here $w = q_1 + q_2$

$q_1 = w - q_2$

Therefore using equations (1) and (2)

$$-\frac{q_1}{w} = \frac{-(w - q_2)}{w} = \frac{-w + q_2}{w}$$

But the problem indicates that $\frac{-q_1}{w} = 7.0$. Therefore

$$\frac{-q_1}{w} = 7.0 = \frac{-w + q_2}{w} = -\frac{w}{w} + \frac{q_2}{w}$$

or $$\frac{-q_1}{w} = -1 + \frac{q_2}{w} = 7.0$$

$$\therefore \quad \frac{q_2}{w} = 7.0 + 1 = 8.0$$

Consequently the heat pump will be 8 times as "efficient" as a 100% efficient resistor. That is it would produce 8 times as much heat as a 100% efficient resistor per caloric of energy input.

Q. 2. For the following reaction at 25°C

$$CuO\,(s) + H_2\,(g) \rightarrow Cu(s) + H_2O(g)$$

Values of S°, the absolute entropies for the substances are:

CuO (s) = 10.4 cal/mole
H_2 (g) = 31.2 cal/mole
Cu (s) = 8.0 cal/g-atm
H_2O (g) = 45.1 cal/mole

Assuming standard conditions find out if the reaction will proceed spontaneously.

Solution : Entropy change is often used to predict the spontaneity of a reaction. A process will occur spontaneously if there is an increase in entropy i.e. ΔS° is positive.

$$\Delta S° = S°_{(products)} - S°_{(reactants)}$$

For the above reaction

$$\Delta S° = S°_{Cu(s)} + S°_{H_2O(g)} - S°_{CuO(s)} - S°_{H_2(g)}$$

$$= (8 + 45.1 - 10.4 - 31.2) \text{cal / deg– mole}$$

$$= +11.5 \text{ cal / deg – mole}$$

ΔS° is positive thus this is a spontaneous reaction.

Q. 3. At 368°K 1 mole of sulfur changes from the monoclinic to the rhombic solid state. ΔH for the transition = –96.01 cal mol^{-1} Calculate the entropy changes if the surroundings is an ice-water bath at 0°C.

Solution : The total change of entropy for the phase transition is given by summing ΔS for the system and surroundigns, thus:

ΔS (universe) = ΔS (system) + ΔS (surroundings)

For a system undergoing transition

$$\Delta S(\text{system}) = \frac{\Delta H}{T}$$

$$= \frac{\left(-96.01\frac{\text{cal}}{\text{mol}}\right)(1.00\text{mol})\left(\frac{4.184\text{J}}{1\text{cal}}\right)}{368°\text{K}}$$

$$= -1.092\frac{\text{Joules}}{°\text{K}}$$

$\Delta S\ (\text{surroundings}) = \frac{q}{T}$ where q = heat transfer. The surrounding is at 0°C = 273°K and the heat transferred to it is the 96.01 cal mol^{-1} given up from the transition

$$\therefore\ \Delta S(\text{surroundings}) = \left(96\frac{\text{cal}}{\text{mol}}\right)(1\text{mol})\left(\frac{4.184\text{J}}{1\text{cal}}\right)\Big/273°\text{K}$$

$$\Delta S(\text{surroundings}) = 1.471\frac{\text{J}}{°\text{K}}$$

$$\therefore\ \Delta S(\text{universe}) = ((-1.092) + 1.471)\frac{\text{J}}{°\text{K}}$$

$$\Delta S(\text{universe}) = 0.379\frac{\text{J}}{°\text{K}}.$$

Q. 4. 1 mole of solid aluminium expands isothermally from 100.0 atm to 1.0 atm. Calculate ΔS (system) for the process given that $\alpha \equiv \frac{1}{V}\left(\frac{\partial V}{\partial T}\right)_P = 69\times10^{-6}\text{K}^{-1}$ and density, d=2.702 × 10^3 kg m^{-3}.

Solution : ΔS for the isothermal expansion of a system is given by

$$\Delta S(\text{system}) = -\int_{P_1}^{P_2}\left(\frac{\partial V}{\partial T}\right)_P dP$$

α = compressibility is given by

$$\alpha = \frac{1}{V}\left(\frac{\partial V}{\partial T}\right)_P$$

$$\therefore \quad \Delta S(\text{system}) = -\alpha V \int_{P_1}^{P_2} dP$$

$$= \alpha V \, (P_2 - P_1)$$

1 mole solid aluminum = 26.9815×10^{-3} kg

$\therefore$ ΔS (system) =

$$-(69\times10^{-6}\,K^{-1})(1\,mol)\left(\frac{26.9815\times10^{-3}\,kg\,mol^{-1}}{2.702\times10^{3}\,kg\,m^{-3}}\right)(1.0-100.0)atm$$

$$\left(101{,}325\ J\,m^{-3}atm^{-1}\right)$$

$$\Delta S\ (\text{system}) = 6.91\times10^{-3}\,\frac{\text{Joules}}{^\circ K}$$

Q. 5. If the pressure and temperature at the center of the earth are 3×10^6 atm and 4000°C respectively, estimate the change in the Gibbs function, ΔG of reaction on going from the crust to the core. Changes in volume, ΔV and entropy, ΔS are 1 $cm^3\ mol^{-1}$ and 2.1 $J\,K^{-1}\,mol^{-1}$ respectively.

Solution : By definition,

$$\left(\frac{\partial \Delta G}{\partial P}\right)_T = \Delta V \quad \text{and} \quad \left(\frac{\partial \Delta G}{\partial T}\right) = -\Delta S.$$

(The parameters are defined in the problem). Using these definitions, an approximation can be made from

$$\Delta G(\text{core}) - \Delta G(\text{crust}) \sim \Delta V(P_{core} - P_{crust}) - \Delta S(T_{core} - T_{crust})$$

where the symbol ~ means approximate or asymptotically equal to.

Since $\Delta V = 1 cm^3\ mol^{-1} = 1 \times 10^{-6}\ m^3\ mol^{-1}$ and the pressure at the center of the earth equals 3×10^6 atm,

$$\Delta V(P_{core} - P_{crust}) \sim (1\times10^{-6}\,m^3mol^{-1})(3\times10^6\times1.013\times10^5\,Nm^{-2})$$

$$= 3.039\times10^5\,J\,mol^{-1}$$

$$\sim 300\ kJ\,mol^{-1}$$

Also,

$$\Delta S(T_{core} - T_{crust}) \sim (2.1 JK^{-1}mol^{-1}) \times (4273\,^\circ K - 298\,^\circ K)$$

$$= 8347.5\ J\,mol^{-1}$$

$$\sim 8.3\ kJ\,mol^{-1}$$

Therefore, ΔG (core)–ΔG (crust) ~ 300 kJ mol^{-1} –8.3 kJ mol^{-1}
= 291.7 kJ mol^{-1}.

Q. 6. What is the amount of work required

(a) to freeze 100 g of water at 0°C when the temperature of the surroundings is 25°C ?

(b) to withdraw the same amount o heat from a body at 10^{-5} °K when the surroundings is 1°K ?

Assume perfect thermodynamic efficiency.

Solution : For a perfect refrigerator operating between the temperatures of T_c and T_h, the

$$COP = \frac{T_c}{T_h - T_c}$$

where COP = coefficient of performance.

The subscripts c and h stand for "cold and hot sinks" respectively.

The amount of work required is given by

$$w = \frac{q}{COP} \qquad (1)$$

where q = amount of heat.

The amount of heat needed in order to freeze water is the number of moles of water multiplied by enthalpy of fusion; given by $n\Delta H_{fusion}$ where n = number of moles. For water

$$\Delta H_{fusion} = 6.01 \text{ kJ mol}^{-1}.$$

Since 100 g of water is to be frozen,

$$n = \frac{\text{weight}}{\text{molecular weight}} = \frac{100\text{g}}{18\text{g mol}^{-1}} = 5.6 \text{ mol}.$$

Therefore, the amount of heat, $q = n\Delta H_{fusion}$.

$$q = (5.6 \text{ mol})(6.01 \text{ kJ mol}^{-1})$$

$$q = 33.6 \text{ kJ}$$

(a) Here T_h = 298°K, T_c = 273°K

$$\therefore \quad COP = \frac{T_c}{T_h - T_c} = \frac{273}{298 - 273} = 10.92.$$

From equation (1), $w = \dfrac{q}{COP} = \dfrac{33.6 \text{ kJ}}{10.92}$

$$\therefore \quad w = 3.07 \text{ kJ}.$$

(b) Here $T_h = 1°K$, $T_c = 10^{-5}\ °K$

$$\therefore \quad COP = \frac{T_c}{T_h - T_c} = \frac{10^{-5}}{1-10^{-5}}.$$

Since the denominator is very close to 1, the COP value is approximately 10^{-5}. Again from equation (1),

$$w = \frac{q}{COP} = \frac{33.6\text{kJ}}{10^{-5}} = 33.6\times10^{5}\,\text{kJ}.$$

Q. 7. For the following reaction at 25°C

$$CuO\ (s) + H_2\ (g) \rightarrow Cu\ (s) + H_2O\ (g)$$

Values of S°, the absolute entropies for the substances are :

CuO (s) = 10.4 cal/mole
H_2 (g) = 31.2 cal/mole
Cu (s) = 8.0 cal/g-atm
H_2O (g) = 45.1 cal/mole

Assuming standard conditions find out if the reaction will proceed spontaneously.

Solution : Entropy change is often used to predict the spontaneity of a reaction. A process will occur spontaneously if there is an increase in entropy i.e., ΔS° is positive.

$$\Delta S° = S°_{\text{(products)}} - S°_{\text{(reactants)}}$$

For the above reaction

$$\Delta S° = S°_{Cu(s)} + S°_{H_2O(g)} - S°_{CuO(s)} - S°_{H_2(g)}$$
$$= (8 + 45.1 - 10.4 - 31.2)\text{cal / deg-mole}$$
$$= +11.5 \text{ cal / deg-mole}$$

ΔS° is positive thus this is a spontaneous reaction.

Q. 8. When mercury is vaporized at its boiling point at standard pressure, the entropy change is 20.7 cal/mol-°K. Determine the boiling point of Hg if the heat of vaporization is 65 cal/g.

Solution : When a process occurs at constant temperature, the change in entropy, ΔS, is equal to the heat absorbed divided by the absolute temperature at which the change occurs, i.e.,

$$\Delta S = \frac{\Delta H}{T},$$

where ΔH is the heat of vaporization (in this case) and T is the absolute temperature. T will be equal to the boiling point of mercury in this problem.

Using this equation, one can solve for T after either converting ΔH to cal/moles or converting ΔS to cal/g. Here, one will convert ΔH from cal/g to cal/mole by multiplying ΔH by the molecular weight of Hg. (MW of Hg = 200.6).

$$\Delta H \text{ in cal/moles} = 65 \text{ cal/g} \times 200.6 \text{ g/mole} = 13039 \text{ cal/mole}$$

One can now solve for T.

$$T = \frac{\Delta H}{\Delta S} \qquad \Delta H = 13039 \text{ cal / mole}$$

$$\Delta S = 20.7 \text{ cal / mole}-{}^\circ K$$

$$T = \frac{13039 \text{ cal / mole}}{20.7 \text{ cal / mole}-{}^\circ K} = 630^\circ K$$

T in °C = 630 – 273 = 357°C.

Q. 9. Determine ΔG° for the reaction

$$4\ NH_3\ (g) + 5O_2\ (g) \rightarrow 4NO\ (g) + 6H_2O\ (\ell)$$

$$\Delta G^\circ_f \text{ of } NH_3\ (g) = -4.0 \text{ Kcal/mole}$$

$$\Delta G^\circ_f \text{ of } NO\ (g) = 20.7 \text{ Kcal/mole}$$

$$\Delta G^\circ_f \text{ of } H_2O\ (\ell) = -56.7 \text{ Kcal/mole}$$

Solution : The change in free energy (ΔG°) may be found by subtracting the sum of free energies (ΔG°_f) of the reactants from the free energies of the products. The free energy of formation of pure elements is always 0. When more than 1 mole of a compound is either reacted or formed, the ΔG°_f of that compound must be multiplied by the stoichiometric coefficient for the specific compound.

$$\Delta G^\circ = (4 \text{ moles} \times \Delta G^\circ_f \text{ of } NO\ (g) + 6 \text{ moles} \times \Delta G^\circ_f \text{ of } H_2O\ (\ell)) - (4 \text{ moles} \times \Delta G^\circ_f \text{ of } NH_3\ (g) + 5 \text{ moles} \times \Delta G^\circ_f \text{ of } O_2)$$

$$G^\circ = (4 \text{ moles} \times 20.7 \text{ kcal/mole} + 6 \text{ moles} \times (-56.7 \text{ kcal/mole})) - (4 \text{ moles} \times (-4.0 \text{ kcal/mole}) + 5 \text{ moles} \times \text{ kcal/mole})$$

$$= -241.4 \text{ kcal.}$$

Q. 10. Determine the entropy difference between solid and liquid states for a substance melting at 100°C and having a heat of fusion of 10,000 J/mol.

Solution : At the melting point, the liquid and solid are in equilibrium, which means the change in free energy (ΔG), i.e., the energy available for useful work, is zero. This is due to the fact that ΔG is defined as the difference of ΔG of the liquid and the ΔG of the solid. Because the system is in equilibrium,

$$\Delta G_{liq} = \Delta G_{solid}.$$

Therefore, ΔG for the whole system is

$$\Delta G_{solid} - \Delta G_{liq} = 0.$$

With this in mind, you can compute the entropy changes (ΔS), i.e., the change in randomness of the system from the equation

$$\Delta G = \Delta H - T\Delta S,$$

where ΔH = enthalpy or heat content and T = temperature in degrees Kelvin (Celsius plus 273°).

For this problem you are given ΔH in the form of heat of fusion, which is the amount of heat necessary to melt one mole of solid. You are also told T. ΔG is equal to zero. Substituting :

$$\Delta G = 0 = \Delta H - T\Delta S \quad \text{or} \quad \Delta S = \frac{\Delta H}{T}$$

$$\Delta S = \frac{10{,}000\ \text{J / mole}}{373°\text{K}} = 26.8\frac{\text{J}}{\text{mole°K}}.$$

4

Third Law of Thermodynamics

Q. 1. A thermodynamic study was made of the propeller-shaped molecule triptycene ($C_{20}H_{14}$; 9, 10-benzeno-9, 10-dihydroanthracen) by measurements of the heat capacity C_p from 10 to 550K. The compound melts at 527.18K with ΔH_m = 7236 cal. mol^{-1}. The molar heat capacities were as follows :

Solid

T.K.	10	20	30	40	50	60	70	80
cal.K^{-1}.mol^{-1}	0.863	4.303	7.731	10.649	13.17	15.40	17.43	19.33
	90	100	120	140	160	180	200	
	21.16	22.98	26.67	30.55	34.63	38.91	43.37	
	220	240	260	280	298.15	320		
	48.01	52.83	57.79	62.88	67.56	73.16		
	350	400	450	500	527.18			
	80.67	92.53	103.85	113.98	119.38			

Liquid

527.18	530	550
130.86	130.90	133.45

Calculate the third-law entropy S for triptycene at 298.15 K and for the liquid at 550 K in units of J.K^{-1}mol^{-1}. 1 cal = 4.1840 J.

Solution: This problem can be solved in a number of steps. The first part, which involves the calculation of the third-law entropy at 298.15°K,

$$S_{298.15} = \Delta S_{0\to 10} + \Delta S_{10\to 298.15},$$

can be solved in two steps and the second part, which involves finding the entropy of the liquid at 550°K,

$$S_{550} = S_{298.15} + \Delta S_{298.15\to527.18} + \Delta S_f + \Delta S_{527.18\to550},$$

where $\Delta S_f = \dfrac{\Delta H_f}{T}$ and ΔH is the change in enthalpy, can be soloved in four step (including the final answer for the first part).

THE PLOT

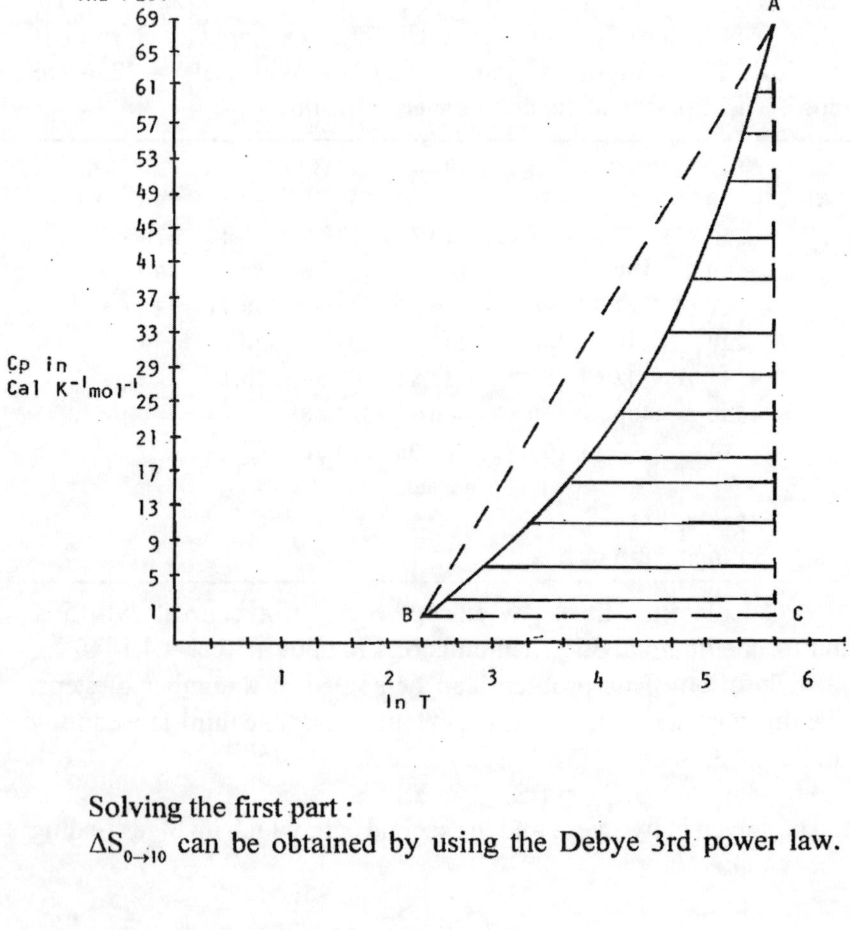

Solving the first part :

$\Delta S_{0\to10}$ can be obtained by using the Debye 3rd power law.

That is, $C_p = AT^3$ where c_p = heat capacity at constant pressure and A = caonstant. By definition,

$$\Delta S = \int_{T_1}^{T_2} \frac{C_p}{T} dt \qquad (1)$$

Substituting the C_p expression into equation (1), it becomes

$$\Delta S = \int_0^{10} \frac{AT^3}{T} dt.$$

$$\Delta S_{0\to10} = \int_0^{10} AT^2 dt = \left.\frac{AT^3}{3}\right|_0^{10} = \left(\frac{AT^3}{3} \text{ at } 10°\text{K}\right) - 0.$$

Note that the answer is 1/3 C_p at 10°K because $AT^3 = C_p°$. The C_p value at 10°K is obtained from the table and it is equal to 0.863.

$\Delta S_{0\to10}$ = 1/3 (0.863) = 0.288 cal $K^{-1}mol^{-1}$. The second step of the first part can be obtained by a graphical integration. That is, plot C_p vs. ln T and $\Delta S_{10\to298.15}$ is the area under the curve. From the table the C_p values and their corresponding lnT values are as follows :

C_p :	0.863	4.303	7.731	10.649	13.17	
ln T :	2.303	2.996	3.401	3.689	3.912	
C_p :	15.40	17.43	19.33	21.16	22.98	26.67
ln T:	4.094	4.248	4.382	4.499	4.605	4.787
C_p :	30.55	34.63	38.91	43.37	48.01	52.83
ln T :	4.942	5.075	5.193	5.298	5.394	5.481
C_p :	57.79	62.88	67.56			
ln T :	5.561	5.635	5.698			

The area under consideration is approximately the shaded portion. If the shaded portion is approximated to be about 2.9/5 of triangle ABC, then its area is 2.9/5 (area of triangle). That is

$$\frac{2.9}{5}[113.2] \approx 65.$$

This is in cal $K^{-1}mol^{-1}$. But 1 cal = 4.1840 J

$$\begin{aligned} S_{298.15} &= \Delta S_{1\to10} + \Delta S_{10\to298.15} \\ &= [0.288 + 65]\ 4.184 \\ &= 272.9 \text{ J K}^{-1}\text{mol}^{-1}. \end{aligned}$$

Solving the sceond part :

$$S_{550} = S_{298.15} + \Delta S_{298.15\to527.18} + \Delta S_f + \Delta S_{527.18\to550}$$

Using the method of graphical integration,

$$\Delta S_{298.15\to527.18} \approx 224 \text{ J K}^{-1}\text{mol}^{-1}$$

and

$$\Delta S_{527.18\to550} \approx 22.2 \text{ J K}^{-1}\text{mol}^{-1}.$$

Now,

$$\Delta S_f = \frac{\Delta H_f}{T} = \frac{(7263)(4.1840)}{527.18}$$

$$\Delta S_f = 57.3 \text{ J K}^{-1}\text{mol}^{-1}$$

$$S_{550} = 272.9 + 224 + 57.3 + 22.2$$
$$= 576.4 \text{ K}^{-1}\text{mol}^{-1}.$$

Q. 2. Between 0 and 60°K the heat capacity of Ag(s) is given approximately by the following expression: $\overline{C}_p$(cal mole^{-1} deg^{-1}) ± 0.1 = –0.023T + 2.5(10^{-3})T^2 – 1.9(10^{-5})T^3. (a) About how much more entropy has a mole of Ag(s) at 60°K than at 0°K? (b) If, according to the thrid law, $S_{Ag(s)}$ is assigned a value of zero at 0°K what is its value at 60°K?

Solution : The increase in entropy of a system due to an increase in temperature can be caluclated since the temperature change can be carried out in a reversible manner. Since the heating is carried out at constant pressure, the heat absorbed in each infinitesimal step is equal to the heat capacity multiplied by the differential increase in temperature, dT. Therefore, the increase in entropy

$$dS = \frac{C_p dT}{T} \tag{1}$$

or

$$\left(\frac{\partial S}{\partial T}\right)_p = \frac{C_p}{T} \text{ where S = entropy.}$$

Integrating equation (1) between T_1 and T_2 gives

$$\int_{S_1}^{S_2} dS = \int_{T_1}^{T_2} \frac{C_p dT}{T} \tag{2}$$

Here S_2 = entropy at 60°K and S_1 = entropy at 0°K.

Therefore,

$$\Delta S = S_2 - S_1 = \int_0^{60} \frac{C_p}{T} dt \qquad (3)$$

Substitute the expression for C_p into equation (3)

$$\Delta S = S_2 - S_1 = \int_0^{60} \left[\frac{-0.023T + 0.0025T^2 - 0.000019T^3}{T} \right] dT$$

$$= -0.023T + \frac{0.0025}{2}T^2 - \frac{0.000019}{3}T^3 \Big|_0^{60}$$

$$= \left[(-0.023)(60) + 0.00125(60)^2 - \frac{0.000019}{3}(60)^3 \right] - 0$$

$$= -1.38 + 4.5 - 1.368$$

$$\Delta S = 1.752 \text{ cal mol}^{-1}\text{K}^{-1}.$$

(b) $\Delta S = S_2 - S_1 = 1.752$. If $S_1 = 0$, then

$$S_2 = 1.752 + 0 = 1.752 \text{ cal mol}^{-1}\text{K}^{-1}.$$

Q. 3. Calculate the standard molar entropy of gaseous platinum at 1000°K. Use only the following data, together with the assumption that Pt is an ideal monatomic gas with negligible electronic excitation below 1000°K.

T, °K	$\frac{\overline{G_T^\circ} - \overline{H_{298}^\circ}}{T}$, cal mole^{-1} °K^{-1}
298.15	–45.96
1000	–49.17

Solution : This problem can be solved in two ways. the first method is by using the expression

$$S_T = \frac{H_T - G_T}{T} = \frac{H_T - H_{298}}{T} - \frac{G_T - H_{298}}{T} \qquad (1)$$

where S = entropy
T = temperature
H = enthalpy
G = Gibb's free enrgy.

The other method is by using

$$S_T = S_{298} + \int_{T_1}^{T_2} \frac{C_p}{T} dT \qquad (2)$$

where C_p = heat capacity at constant pressure. For an ideal monatomic gas,

$$C_P = C_v + R = \frac{5}{2}R.$$

C_V= heat capacity at constant volume.
R = gas constant. Using equation (1)

$$S_T = \frac{H_T - G_T}{T} = \frac{H_T - H_{298}}{T} - \frac{G_T - H_{298}}{T}$$

$$= \frac{H_{1000} - H_{298}}{1000} - \frac{G_{1000} - H_{298}}{1000} \qquad (3)$$

At 1000°K, $\frac{G_{1000} - H_{298}}{1000}$ is given to be –49.17 cal mol⁻¹K⁻¹,

and

$$H_{1000} - H_{298} = \Delta H = \int_{298}^{1000} C_p dT$$

$$= \int_{298}^{1000} \frac{5}{2} R\, dT = \frac{5}{2} R(1000 - 298) = \frac{5}{2} R(702)$$

$$H_{1000} - H_{298} = \frac{5}{2}(1.987)(702) = 3487.2 \text{ cal mol}^{-1}.$$

Therefore equation (3) becomes

$$S_{1000} = \frac{3487.2}{1000} - (-49.17) = 52.7 \text{ cal mol}^{-1}\text{K}^{-1}.$$

From equation (2) S_{298} is given to be +45.96 cal $mol^{-1}K^{-1}$.

Note that $S_{298} = -\frac{G_{298} - H_{298}}{298} = 45.96$. Therefore,

$$S_{1000} = S_{298} + \int_{298}^{1000} \frac{C_P}{T} dT$$

$$= 45.96 + \frac{5}{2} R[\ln 1000 - \ln 298]$$

$$= 45.96 + \frac{5}{2} R[6.91 - 5.697]$$

$$= 45.96 + 6.02$$

$$= 51.98 \approx 52 \text{ cal mol}^{-1}\text{K}^{-1}$$

The two methods yield acceptable answers.

Q. 4. What is the standard free energy for the oxidation of glucose given by the reaction

$$C_6H_{12}O_6\ (s) + 6O_2\ (g) = 6O_2\ (g) + 6H_2O\ (\ell)\ ?$$

The standard free energy of formation of $C_6H_{12}O_6$ (s) = –218 kcal mole^{-1}, for CO_2 (g), it is –94.0 kcal mole^{-1} and for H_2O (l), it is –57.0 kcal mole^{-1}.

Solution : The standard free energy of reaction ΔG° is given by

$$\Delta G^\circ = \Sigma \Delta G^\circ_{f(products)} - \Sigma \Delta G^\circ_{f(reactants)}$$

where ΔG°_f is the standard free energy of formation. The standard heat of formation of an element in its standard state is zero.

$$\Delta G^\circ = [6(-94.0) + 6(57.0)] - [-218 + 0]$$

$$\Delta G^\circ = -688 \text{ kcal mole}^{-1}.$$

Q. 5. By measurement of the equilibrium constant for the reaction $H_2O(liq) + CO(g) \rightleftarrows HCOOH(liq)$, the standard free energy of formation of formic acid from the elements is found to be –85,200 cal mole^{-1} at 298°K. The standard free energy of formation of formic acid from the elements at 298°K is –85,370 cal mole^{-1} by heat capacity measurements. How does the agreement between the two results offer experimental evidence for the validity of the third law of thermodynamics ?

Solution : Both methods give results that are very close.

In the former method, the calculation of ΔG was not based on the assumption of the validity of the third law. However, in the latter case the third law was assumed to be valid. (That is, the calculation of ΔG was from the heat capacities).

The agreement between the two results is therefore an evidence that the third is valid.

Q. 6. Use the following table to calculate the enthalpy of propane at 25°C and at one atm assuming the enthalpy of solid carbon and hydrogen gas to be zero that temperature and pressure.

Bond	Energy	Bond	Energy
H-H	104.2	H-1	71.4
C-C	83.1	C-N	69.7
Cl-Cl	58.0	C-O	84.0

Br-Br	46.1	C-Cl	78.5
I-I	36.1	C-Br	65.9
C-H	98.8	C-I	57.4
N-H	93.4	O-O	33.2
O-H	110.6	N≡N	226
H-Cl	103.2	C=C	147
H-Br	87.5	C≡C	194

C=O 164 in formaldehyde
174 in other aldehydes
174 in ketones,

Resonance energy in kcal/g mole
Benzen ring = 37
Naphthalene = 75
Carboxylic acids = 28
Esters = 24

The heat of vaporization for carbon (s) to carbon (g) = 171.70 kcal/mole.

Solution : The formation reaction of propane from carbon and hydrogen is

$$3C\ (s) + 4H_2\ (g) = C_3H_8\ (g).$$

The subscripts (s) and (g) represent solid and gas respectively.

Since the formation of propaten takes place at constant pressure, the heat required would be the difference between that of the product and those of the reactants.

From the reaction, three carbons are vaporized and four H-H bonds are broken while two C-C bonds and eight C-H bonds are formed.

In the vaporization of a carbon atom, 171.70 kcal is required. Therefore, the vaporization of three carbon atoms would require 3 (17.70) = 515.10 kcal.

From the table, the breaking of four H-H bonds would require 4(104.2) = 416.8 kcal. Therefore, total energy added = 931.9 kcal.

The energy released in the formation of the C_3H_8 molecule is

$$2\ (83.1) + 8\ (98.8) = 956.6 \text{ kcal.}$$

Hence the net energy = 931.9 – 956.6 = 24.7 kcal. Note that since the elements (C and H) were at the reference state, the enthalpy would be –24.7 kcal at 25°C.

Q. 7. Estimate the zero-point entropy of NO.

Solution : The solid state NO can exist in two orientations which have practically equal chances of existence. The orientations are NO NO and NO ON.

Consequently, the zero-point entropy = k 1n 2^N where k = Boltzmann's constant and N = Avogadro's number.

Zero point entropy = (1.987) (0.693)
= 1.38 cal mol^{-1} K^{-1}.

5

Statistical Thermodynamics

Q. 1. Data for monatomic fluorine at 1000 °K are as follows :

Term symbol	Wave numbers $\bar{v} = \varepsilon / hc / (cm^{-1})$	Electronic degeneracy, gel
$P_{3/2}$	0.0	4
$P_{1/2}$	404.0	2
$P_{5/2}$	102, 406.5	6

Calculate the fraction of fluorine atoms in each of the first three electronic levels.

Solution : The electronic partition function of the fluorine atom is calculated using the relation:

$$Q_{elec} = \Sigma g_j e^{-\varepsilon_{elec,j/kT}}$$

$$= g_0 + g_1 e^{-\varepsilon_{elec,1/kT}} + g_2 e^{-\varepsilon_{elec,2/kT}} + ...$$

where $\varepsilon_{elec,j}$ is the electronic energy of level j taken relative to the ground state, and g_j is the degeneracy number. Using the given data,

$$Q_{elec.} = g_0 e^{-\varepsilon_{0,el/kT}} + g_1 e^{-\varepsilon_{1,el/kT}} + g_2 e^{-\varepsilon_{2,el/kT}} + ...$$

$$= 4e^{-(\varepsilon_{0,el}/hc)(hc/kT)} + 2e^{-(\varepsilon_{1,el}/hc)(hc/kT)} + 6e^{-(\varepsilon_{2,el}/hc)(hc/kT)} + ...$$

Computing the hc/kT term gives

$$\frac{hc}{kT} = \frac{(6.63\times10^{-34}\,J\,s)(3.0\times10^{10}\,cm\,s^{-1})}{(1.38\times10^{-23}\,JK^{-1})(1000°\,K)}$$

$$= 1.44\times10^{-3}\,cm$$

From the table in the problem set, $\varepsilon_{0,el}$ / hc = 0.0, $\varepsilon_{1,el}$ / hc = 404 cm^{-1} and $\varepsilon_{2,el}$ / hc = 102,406.5 cm^{-1}. Therefore,

$$Q_{elec} = 4e^0 + 2e^{-(404cm^{-1})\,(1.44\times10^{-3}cm)}$$
$$+\ 6e^{-(102,406.5cm^{-1})(1.44\times10^{-3}cm)}$$
$$= 4e^0 + 2e^{-0.5817} + 6e^{-147.47}$$
$$= 4 + 1.117 + 5.404\times10^{-64}$$
$$\simeq 5.117.$$

The Maxwell-Boltzman distirbution law is used to calculate the fraction of fluorine atoms in each of the first three electronic levels. This law is given by

$$\frac{n_j}{N} = \frac{g_{elec,j}e^{-\varepsilon_{elec,j}/kT}}{Q_{elec}} \qquad (2)$$

For the ground level, equation (2) is

$$\frac{n_0}{N} = \frac{g_{elec,0}e^{-\varepsilon_{elec,o}/kT}}{Q_{elec}}$$
$$= \frac{4e^0}{5.117}$$
$$= \frac{4}{5.117}$$
$$= 0.782$$

For the first level

$$\frac{n_1}{N} = \frac{g_{elec,1}e^{-(\varepsilon_{elec,1}/hc)(hc/kT)}}{Q_{elec}}$$
$$= \frac{2e^{-(404/1.44\times10^{-3})}}{5.117}$$
$$= \frac{2e^{-0.5817}}{5.117}$$
$$= \frac{2(0.558)}{5.117}$$
$$= 0.218.$$

For the second electronic level

$$\frac{n_2}{N} = \frac{g_{elec,2}e^{-(\varepsilon_{elec,2}/hc)(hc/kT)}}{Q_{elec}}$$

$$= \frac{6e^{-(102,406.5\text{cm}^{-1})(1.44\times10^{-3}\text{ cm})}}{Q_{elec}}$$

$$= \frac{6e^{-147.47}}{5.117} \simeq 0$$

Apparently, the fraction of fluorine atoms in the second electronic level is very small and so can be neglected, by approximating it to be zero.

Q. 2. Calculate the molecular translational partition functions in a volume of 1 cm³ at 298°K for the following molecules (a) H_2, (b) CH_4, (C) C_8H_{18}.

Solution : The molecular translational partition function is given by

$$q_{trans.} = \left[\frac{2\pi mkT}{h^2}\right]^{3/2} V$$

where h is Planck's constant, m is the mass of the molecule and V is the volume. Taking the mass out of the square bracket, gives

$$q_{trans.} = \left[\frac{2\pi kT}{h^2}\right]^{3/2} Vm^{3/2}$$

$$= \left[\frac{2\pi(1.38\times10^{-23}\text{ JK}^{-1})(298°\text{ K})}{(6.63\times10^{-34}\text{ J Sec})^2}\right]^{3/2} (1\text{ cm}^3)\left(\frac{1\text{m}}{100\text{ cm}}\right)^3 \text{m}^{3/2}$$

$$= 1.42\times10^{64}(\text{J}^{-1}\text{s}^{-2})^{3/2}\text{ m}^{3/2}\text{ (meters)}^3 \quad (1)$$

(a) For H_2 :

$$\text{Mass per molecule of } H_2 = \frac{(2.016\text{ g mol}^{-1})\left(\frac{1\text{kg}}{1000\text{g}}\right)}{6.02\times10^{23}\text{ mol}^{-1}}$$

$$= 3.348\times10^{-27}\text{ kg}$$

From equation (1)

$$q_{trans.} = 1.42\times10^{64}(J^{-1}S^{-2})^{3/2}(3.348\times10^{-27}\,kg)^{3/2}$$
$$= 2.750\times10^{24}(J^{-1}S^{-2})^{3/2}\,kg^{3/2}(met)^3$$
$$= 2.750\times10^{24}\left(\frac{sec^2}{kg(met)^2}\frac{1}{sec^2}\right)^{3/2} kg^{3/2}(met)^3$$
$$= 2.750\times10^{24}\,kg^{-2/3}met^{-3}kg^{3/2}met^3$$
$$= 2.750\times10^{24}$$

(b) For CH_4 :

$$\text{mass per molecule of } CH_4 = \frac{(16.04g\ mol^{-1})\left(\frac{1kg}{1000g}\right)}{6.02\times10^{23}\,mol^{-1}}$$
$$= 2.664\times10^{-26}\,kg$$

$$\therefore\ q_{trans.} = (1.42\times10^{64})(2.664\times10^{-26})^{3/2}$$
$$= 6.175\times10^{25}$$

(c) For C_8H_{18} :

$$\text{Mass per molecule of } C_8H_{18} = \frac{(114.22g\ mol^{-1})\left(\frac{1kg}{1000g}\right)}{6.02\times10^{23}\,mol^{-1}}$$
$$= 1.897\times10^{-25}\,kg$$

$$\therefore\ q_{trans.} = (1.42\times10^{64})(1.897\times10^{-25})^{3/2}$$
$$=1.173\times10^{27}$$

Q. 3. Two molecules $^{14}N_2$ and $^{14}N^{15}N$ have the same internuclear distance given as 0.1095 nm. Evaluate the molar rotational partition function at 298°K for both molecules.

Solution : The molar partition function, Q and the molecular partition function, q are related to each other by the equation

$$Q = q^N \tag{1}$$

where N is the Avogadro number. Taking the ℓn of both sides changes equation (1) to

$$\ell n Q = N\ \ell n\ q \tag{2}$$

The molecular rotational partition function is given as

$$q_{rot.} = \frac{8\pi^2 IkT}{\sigma h^2} \qquad (3)$$

where I is the moment of inertia about the center of mass, h is the Planck's constant and σ, the symmetry number, is equal to the number of indistinguishable positions into which the molecule can be turned by simple rigid rotation. σ takes the values 1 for heteronuclear diatomics and 2 for homonulcear diatomics.

The moment of inertia is calculated using the internuclear distance given by the relation

$$I = \mu r^2$$

where r is the internuclear distance and μ is the reduced mass given as

$$\mu = \frac{m_1 m_2}{m_1 + m_2}$$

m_1 and m_2 are the masses of atoms 1 and 2 of the molecule. Therefore,

$$I = \frac{m_1 m_2}{m_1 + m_2} r^2.$$

(a) $^{14}N_2$: Here m_1 and m_2 are the same atoms. Therefore,

$$I = \frac{m^2}{2m} r^2$$

$$= \frac{\left[(14.007\text{g mol}^{-1})\left(\frac{1\text{kg}}{1000\text{kg}}\right)\right]^2 \left[0.1095\text{nm}\left(\frac{10^{-9}\text{ m}}{1\text{nm}}\right)\right]^2}{2(14.007\text{g mol}^{-1}), \left(\frac{1\text{kg}}{1000\text{g}}\right)(6.02\times10^{23}\text{ mol}^{-1})}$$

$$= 1.393\times10^{-46}\text{ kg.m}^2.$$

The symmetry number, σ for $^{14}N_2$ = 2. Therefore, using equation (3) gives

$$q_{rot.} = \frac{8\pi^2(1.393\times10^{-46}\text{ kg.m}^2)(1.38\times10^{-23}\text{ J K}^{-1})(298\ ^\circ\text{K})}{2(6.63\times10^{-34}\text{ J S})^2}$$

$$= 51.44\frac{\text{J kg m}^2}{\text{J}^2\text{ sec}^2} = 51.44\frac{\text{J}^2}{\text{J}^2}$$

$$= 51.44.$$

The molar rotational partition function is thus given by

$$\ell n\, Q_{rot.} = N\, \ell n\, 51.44$$

$$= (6.02 \times 10^{23}\, mol^{-1})\, \ell n\, 51.44$$

$$= 2.37 \times 10^{24}$$

$$\therefore \quad Q_{rot.} = e^{2.37 \times 10^{24}}$$

(b) For $^{14}N^{15}N$,

$$I = \frac{m_1 m_2}{m_1 + m_2} r^2 = \mu r^2$$

$$= \frac{(14.007 \times 10^{-3}\, kg\, mol^{-1})(15.004 \times 10^{-3}\, kg\, mol^{-1})(0.1095 \times 10^{-9}\, m)^2}{(29.011 \times 10^{-3})(6.02 \times 10^{23})}$$

$$= 1.44 \times 10^{-46}\, kg\, m^2$$

For this molecule, $\sigma = 1$

$$\therefore \quad q_{rot.} = \frac{8\pi^2(1.44 \times 10^{-46}\, kg\, m^2)(1.38 \times 10^{-23}\, J\, K^{-1})(298\ ^\circ K)}{(1)(6.63 \times 10^{-34}\, J\, S)^2}$$

$$= 106.37$$

The molar rotational partition function is thus

$$\ell n\, Q_{rot.} = N\, \ell n\, 106.37$$

$$= (6.02 \times 10^{23}\, mol^{-1})\, \ell n\, 106.37$$

$$= 2.81(10^{24})$$

and

$$Q = e^{2.81 \times 10^{24}}$$

Q. 4. the degeneracies of the singlet (paired electron spins) or triplet (unpaired spins) levels of a certain molecule are 1 and 3, respectively. The energy of the singlet state is greater than that of the triplet by a factor ε. (a) Write the electronic factor in the partition function of the molecule. Indicate the approximations made. (b) Given that the factor $\varepsilon = 1.38 \times 10^{-14}$ erg and $T = 100°K$, calculate the ratio of the population of the singlet level to the population of the triplet level.

Solution : (a) The electronic partition function of the molecule is given by the relation

$$Q_{elec} = \sum_i g_i e^{-\varepsilon_{elec,j}/kT}$$

$$= g_o + g_1 e^{-\varepsilon_{elec,1}/kT} + g_2 e^{-\varepsilon_{elec,2,}/kT} + \ldots$$

For every level i with $\varepsilon_i > \varepsilon$, assume that $g_i e^{-\varepsilon_i/kT} << e^{-\varepsilon/kT}$,

$$\therefore \quad Q_{elec} = 3 + e^{-\varepsilon/kT}$$

(b) The ratio of the population of the singlet level to the population of the triplet level is given by

$$\frac{N_{singlet}}{N_{triplet}} = \frac{e^{-\varepsilon/kT}}{3}$$

$$= \frac{1}{3}\exp\left\{\frac{-1.38\times10^{-14}\,\text{erg}}{(1.38\times10^{-16}\,\text{erg }^\circ\text{K}^{-1})(100\text{K})}\right\}$$

$$= \frac{1}{3}e^{-1}$$

$$= \frac{1}{3}(0.3678)$$

$$= 0.123$$

Q. 5. In the vapor state the atoms of thallium have a doublet ground state of energy separation $S_{1/2} - S_{3/2}$ equal to $\varepsilon/k = 11{,}200\ ^\circ$K. Calculate the entropy of thallium vapor at 877 °K and 1 atm. Compare it with the value of 203 ± 1 J mole^{-1}K^{-1} obtained by integrating the calorimetric data and reducing the value to the standard atomspheric pressure. $\omega_{e_1} = 2$ and $\omega_{e_2} = 4$.

Solution : If N molecules in a gas could be identified, they could be arranged N! ways between the N different microstates that they happen to occupy. Therefore, the canonical ensemble parition function calculated, which assumes that the molecules are distinguishable, yields too large a value by a factor of N!. Consequently, for nonlocalized subsystems,

$$Q = \frac{1}{N!}q^N$$

where Q = molar partition function and $q = q_{trans}q_{elec.}$ = individual molecular partition function. By definition,

$$q_{trans} = \frac{V}{\Lambda^3}$$

where $$\Lambda \quad \frac{h}{(2\pi mkT)^{1/2}}$$

$$= \frac{(6.63\times10^{-27}\,\text{erg sec})}{\left[2\pi\left(\frac{204.37\text{g mole}^{-1}}{6.02\times10^{23}\,\text{mole}^{-1}}\right)(1.38\times10^{-16}\,\text{erg °K}^{-1})(877\ \text{°k})\right]^{1/2}}$$

$$= 4.124\times10^{-10}\,\text{cm.}$$

From the ideal gas law

$$V = \frac{RT}{P}$$

$$= \frac{(8.314\ \text{Nm °K}^{-1}\ \text{mole}^{-1})(298\ \text{°K})}{101325\text{Nm}^{-2}}\left(\frac{100\text{cm}}{1\text{m}}\right)^3$$

$$= 7.198\times10^4\,\text{cm}^3\,/\,\text{mole}$$

$$q_{elec} = \omega_{e_1} + \omega_{e_2}e^{-\varepsilon_{e_2}/kT}$$

where ω_e = the degeneracy of the electronic ground state.

$$q_{elec} = 2 + 4e^{-11200/877}$$
$$= 2 + 0.00001137$$
$$= 2.$$

The expression for the molar entropy is given by

$$S = Nk\,\ell n\left[\frac{Ve^{5/2}\Omega e_1}{\wedge^3 N}\right]$$

where $\Omega e_1 = 2 =$ the first fold degenerate of the electronic ground state, normally the only electronic state thermally accessible.

$$S = 8.317\times10^7\,\frac{\text{erg}}{\text{mole° K}}\times$$

$$\ell n\left[\frac{(7.198\times10^4\,\text{cm}^3\text{mole}^{-1})e^{5/2}(2)}{(4.124\times10^{-10}\,\text{cm})^3(6.02\times10^{23}\,\text{mole}^{-1})}\right]$$

$$= 2.03\times10^9\,\text{erg mole}^{-1}\,\text{°K}^{-1}$$

$$= 203.3\ \text{J mole}^{-1}\,\text{°K}^{-1}$$

This result is within the experimental error of the calorimetric data (203 ± 1 J mole^{-1} °K^{-1}).

Q. 6. The rotational spectrum of HF has lines 41.9 cm^{-1} apart. Calculate the moment of inertia and the bond length of this molecule.

Solution : The rotatioanl constant of the molecule is given by

$$B = \frac{h}{8\pi^2 I} \qquad (1)$$

where h = Planck's constant.

I = moment or inertia.

The spacing between the lines is given to be 41.9 cm^{-1}. This is also equal to 2B. Therefore,

$$B = \frac{41.9}{2} = 20.95\text{cm}^{-1}$$

$$= 20.95\text{cm}^{-1} \times 2.99 \times 10^{10} \text{cm sec}^{-1}$$

$$= 6.281 \times 10^{11} \text{sec}^{-1}.$$

Rearranging equation (1) to solve for I yields,

$$I = \frac{h}{8\pi^2 B} = \frac{(6.626 \times 10^{-34} \text{J sec})}{8\pi^2 (6.281 \times 10^{11} \text{sec}^{-1})}$$

$$= 1.336 \times 10^{-47} \text{kg m}^2$$

The bond length is given by

$$r = \left(\frac{I}{\mu}\right)^{1/2}$$

where

$$\mu = \frac{M_H M_F}{M_{HF} L}.$$

Here M = molecular weight (atomic wt.)

L = Avogadro's number.

$$\text{Hence, } \mu = \frac{(1)\text{g mole}^{-1}(19)\text{g mole}^{-1}}{(20)\text{g mole}^{-1}(6.02 \times 10^{23} \text{mole}^{-1})} = 0.158 \times 10^{-23} \text{g}$$

or $\mu = 0.158 \times 10^{-26}$ kg.

Using the calculated values in equation (2) yields

$$r = \left(\frac{1.336 \times 10^{-47} \text{kg m}^2}{0.158 \times 10^{-26} \text{kg}}\right)^{1/2}$$

$$= (8.456 \times 10^{-21} \text{m}^2)^{1/2}$$

$$= 9.196 \times 10^{-11} \text{m} \times 10^9 \text{nm / m}$$

$$= 9.196 \times 10^{-2} \text{nm}$$

Q. 7. Find the difference in characteristic vibrational frequency of a diatomic molecule of mass 20 amu if (a) the mass is equally distributed between the two ends, and (b) one atom of 1 amu is at one end and an atom of 19 amu is at the other.

Solution : The vibrational frequency is frequency of the oscillation between the two atoms of the molecule. It is expressed quantitatively by

$$v = ½\pi\sqrt{K/M_{eff}}$$

where v is the vibrational frequency, K a force constant, and M_{eff} the effective mass of the molecule.

$$M_{eff} = \frac{M_A M_B}{M_A + M_B}$$

where M_A is the mass of one atom and M_B is the other.

(a) If the mass is equally distributed between both atoms, each weighs 10 amu.

$$M_{eff} = \frac{10\times10}{10+10} = 5$$

Assume that K = 1.

$$v = \frac{1}{2\pi}\sqrt{\frac{1}{5}} = 0.0711$$

(b) $$M_{eff} = \frac{1\times19}{20} = \frac{19}{20}$$

$$v = \frac{1}{2\pi}\sqrt{\frac{19}{20}} = 0.1632$$

Therefore, the atoms in the molecule described in b are vibrating more quickly than those in molecule a.

Q. 8. Calculate the characteristic vibrational frequency of the lowest vibration state for H_2. Assume k = 5.1 newtons/cm and h = 6.63×10^{-34} J-sec.

Solution : Picture a diatomic molecule, such as H_2, as two balls connected by a spring. The spring represents the chemical bond. When the two balls (atoms) are pulled apart the string acts as a restoring force to bring them back together to the equilibrium separation. When they are compressed, the string tends to force them apart. Such a systm, in which the restoring force is proportional to the amount of distortion, is called a harmonic oscillator. The vibrational frequency of the spring going back and forth is given by the formula

$$v = \frac{1}{2\pi}\sqrt{k / M_{eff}}$$

where v = characteristic vibrational frequency of the particular bond, k = force constant and M_{eff} = effective mass for the vibrating atoms :

$$M_{eff} = \frac{M_A M_B}{M_A + M_B}$$

If the masses of the two atoms are equal, M_{eff} becomes $\frac{M^2}{2M} = ½M$.

To find the characteristic vibration frequency, use this formula.

To find the M_{eff}, find the mass of one hydrogen atom. The number of grams in one mole (molecular weight) is one. There are 6.02×10^{23} atoms of H in one mole. Thus, the mass of one atom is

$$\frac{1g / mole}{6.02 \times 10^{23}} = 1.66 \times 10^{-24} \text{ g.}$$

Thus,

$$M_{eff} = \frac{(1.66 \times 10^{-24} \text{ g})}{2} = 8.31 \times 10^{-25} \text{ g.}$$

Solving for v :

$$k = 5.1 \frac{\text{newtons}}{\text{cm}} = 5.1 \frac{\text{kgM}}{\text{cm.sec}^2}$$

$$v = \frac{1}{2\pi}\sqrt{\frac{5.1 \frac{\text{kgM}}{\text{cm.sec}^2} \times 1000 \frac{\text{g}}{\text{kg}} \times 100 \frac{\text{cm}}{\text{m}}}{8.31 \times 10^{-25} \text{ g}}}$$

$= 1.25 \times 1014 \text{ sec}^{-1}$.

Note : 1000 g/kg and 100 cm/m are conversion factors used to obtain the correct units.

6

Liquids and Solutions

Q. 1 : 13.00g NaOH (m. w. = 40.01 g mol^{-1}) and 87.00g H_2O are mixed to form a solution of density 1.1421×10^3 kg m^{-3}. Calculate a) wt % of NaOH, (b) wt % of H_2O, (c) the mass of the solvent, (d) the mole fraction of NaOH, (e) the mole fractions of H_2O, (f) (g solute) (dm^3 soln)$^{-1}$, (g) the molarity of the solution.

Solution: NaOH is the solute and H_2O is the solvent. By definition wt % of a solute in solution is given by

$$\frac{\text{grams solute}}{\text{grams solute } + \text{grams solvent}} \times 100\%$$

$$\text{(a) Wt\% of NaOH} = \frac{(\text{g NaOH})}{(\text{g NaOH}) + (\text{g } H_2O)} \times 100\%$$

$$= \frac{13.00\text{g}}{(13.00\text{g}) + (87.00\text{g})} \times 100\%$$

$$= 13.00\%$$

(b) For wt % water, there is 100g of solution. So wt % water = 100.00 – 13.00 = 87.00%

(c) The number of moles of solute is given by

$$n_{NooH} = \frac{13.00\text{g}}{(40.01\text{g mol}^{-1})} - 0.325 \text{ mole}$$

The number of moles of solvent is given by

$$n_{H_2O} = \frac{87.00\text{g}}{(18.015\text{g mol}^{-1})} = 4.83 \text{ mol}$$

Molatity of the solution is defined as

$$\frac{\text{moles of solute}}{\text{kilograms of solvent}}$$

$$\text{Molatity} = \frac{0.325\text{ mol}}{(87.00\text{g } H_2O)\left(\frac{1\text{kg } H_2O}{1000\text{g}}\right)} = 3.73\text{m}$$

Since molatity depends on moles of solute rather than grams, 13.00g NaOH is converted to moles. The molal concentration also gives an equivalence that will be useful in finding the mass of the solvent. The mass of the solvent is

$$13.00\text{g NaOH} \times \frac{1\text{mole NaOH}}{40.01\text{g NaOH}} \times \frac{1\text{kg } H_2O}{3.73\text{ moles NaOH}}$$

$$= 8.700 \times 10^{-2}\text{ kg } H_2O$$

(d) Mole fraction, X is

$$\frac{\text{mole solute}}{\text{moles solute + moles solvent}} \text{ by definition.}$$

$$\text{Therefore,} \quad X_{NaOH} = \frac{0.325}{0.325 + 4.83} = 0.0631$$

The mole fractions of the solute and the solvent must sum up to 1. Thus the mole fraction

(e) $X_{H2O} = 1.0000 - 0.0631 = 0.9369$

or it can be obtained by using the difinition of mole fraction. Therefore

$$X_{H_2O} = \frac{4.83}{4.83 + 0.325} = \frac{4.83}{5.155} = 0.9369$$

(f) The volume of solution, V, using 100.00g basis is calculated using the fact that

$$\text{Volume} = \text{mass} \div \text{density}$$

$$\therefore \quad V = \frac{(100.00\text{g})\frac{10^{-3}\text{ kg}}{1\text{g}}}{1.1421 \times 10^3\text{ kg m}^{-3}} = 87.56 \times 10^{-6}\text{ m}^3$$

$= 87.56 \times 10^{-3}\text{dm}^3$ or 87.56×10^{-3} liters

$\therefore$ (g solute $(\text{dm}^3\text{ soln})^{-1}$ = 13.00g NaOH) $(87.56 \times 10^{-3}\text{dm}^3)^{-1}$

$= 148.5\text{g NaOH } (\text{dm}^3\text{ soln})^{-1}$

(g) Molarity by definition is

$$C = \frac{\text{moles solute}}{\text{liters solution}} = \frac{0.325 \text{ mol NaOH}}{87.56 \times 10^{-3} \text{ dm}^3} = 3.71\text{M}.$$

Q. 2 : Calculate the vapor pressure of a solution of 20.0 grams of sucrose, $C_{12}H_{22}O_{11}$, in 40.0 grams of water at 34°C. The vapor pressure of water is 40.00 mm at 34°C.

Solution: Raoult's law states that at a given temperature, the partical pressure P_i of a component in solution is proportional to the mole fraction X_i as follows:

$$P_i = X_i P_i^o \qquad (1)$$

where P_i^o is the vapor pressure of pure component i.

Since the equation is not directly related to the weights of the substances, it is necessary to convert them (the weights) to moles by the relation,

$$n = \frac{\text{weight}}{\text{molecular weight}} \qquad (2)$$

where n = number of moles. The molecular weights of $C_{12}H_{22}O_{11}$ and H_2O are 342 g $C_{12}H_{22}O_{11}$/mole and 18 g H_2O/mole respectively. Therefore, their respective mole are

$$(20\text{g } C_{12}H_{22}O_{11})\left[\frac{1 \text{ mole } C_{12}H_{22}O_{11}}{342\text{g } C_{12}H_2O_{11}}\right]$$

$$= 0.0584 \text{ mole } C_{12}H_{22}O_{11}$$

and

$$(40\text{g } H_2O)\left[\frac{1 \text{ mole } H_2O}{18.0\text{g } H_2O}\right]$$

$$= 2.22 \text{ moles } H_2O.$$

Total moles in the solution = (0.0584 + 2.22) moles

= 2.28 moles

The mole fraction of water in the solution is

$$X_{H_2O} = \frac{{}^{n}H_2O}{{}^{n}\text{total}}$$

where x = mole fraction,

$$X_{H_2O} = \left[\frac{2.22 \text{ moles}}{2.28 \text{ moles}}\right]$$

$$X_{H_2O} = 0.974$$

Using Raoult's law to calculate the vapor pressure of the solution, equation (1) becomes

$$P_{sol} = X_{H_2O} P^{O}_{H_2O}$$

$$= (0.974)\ (40.00\text{mm})$$

$$P_{solution} = 39.0 \text{ mm.}$$

Q. 3 : At 25°C, the vapor pressure of water is increased from 30 to 31 mm Hg against an external pressure of 10 atm. If the vapor pressure is to be maintained at 30 mm Hg for 1 liter (55.5 moles) of water at 25°C and 10 atm pressure, (a) How much NaCl is to be added to the water ? (b) What is the osmotic pressure of this solution if NaCl forms an ideal solution with the water ?

Solution : (a) Here it could be assumed that the vapor pressure of the solution is that of the solvent because the vapor pressure of the solute can be assumed to be very negligible. Since this is an ideal solution, the vapor pressure of the solvents is given by Raoult's law

$$P_1 = X_1 P_1^O$$

$$P_1 = (1 - X_2)\ P_1^O \qquad (1)$$

where subscript 1 indicates the solvent and subscript 2 indicates the solute.

From Equation (1) the effective mole fraction of the solute NaCl can be written as

$$X_2 = \frac{P_1^O - P_1}{P_1^O}$$

$$\therefore \qquad X_2 = \frac{1}{31}$$

$$= 0.032$$

When NaCl dissociates in the solution it forms ions, Na^+ and Cl^- per formula weight. Thus the actual mole fraction of NaCl must be $\frac{0.032}{2} = 0.016$. The moles of solute (NaCl) to be added to 55.5 moles of water are given by

$$X_2 = \frac{n_2}{n_1 + n_2} \qquad (2)$$

where n_2 = moles of NaCl
n_1 = moles of H_2O

$\therefore$ From Equation (2)

$$0.016 = \frac{n_2}{(55.5 \text{ moles} + n_2)}$$

$$0.888 + 0.016\ n_2 = n_2$$

$$n_2 = .904$$

(b) The osmotic pressure is defined as the pressure difference across a membrane required to prevent flow across the membrane in eigher side. In this problem there is an external pressure of 10 atm and to maintain the vapor pressure at 30 mm Hg, the 10 atm pressure must be constant. Thus the osmotic pressure is 10 atm.

Q. 4 : Consider the process

$$H_2O \text{ (solution, 25°C)} = H_2O \text{ (pure, 25°C)}.$$

If the solution is NaCl of concentration m and has an osmotic pressure of 2.0 atm at 25°C, calculate ΔG_{298}, the free energy change for the process.

Solution : Here the NaCl is the solute because its mole fraction can be varied. For the above process, we consider the solution to be separated from a small amount of the pure solvent at the same temperature by means of a semipermeable membrane which is permeable only to solvent molecules. Then the excess pressure which must be placed on the solution to prevent any diffusion of solvent through the membrane is known as the Osmotic pressure. The diffusion arises because the solvent is at a lower chemical potential in the solution than it is in the pure liquid.

Let P be the pressure on the pure solvent and P be the pressure on the solution. Then at equilibrium the Osmotic pressure

$$\pi \equiv P' - P.$$

For the process

$$H_2O \text{ (solution)}, \ G = G_P^o, + RT \ln P$$

and for

H_2O (pure), $G = G^o_{P'} + RT\,\ell n\,P^o$

At equilibrium

$$G_P + RT\,\ell n\,P^o = G^o_{P'} + RT\,\ell n\,P$$

$$\Delta G = RT\,\ell n\,(P^o/P) = G^o_P - G^o_{P'}$$

$$= \int_P^{P'} \overline{V}\,dP = \overline{V}\,(P' - P)$$

where $\overline{V}$ is the molar volume of the pure solvent. To a reasonably great extent, V remains constant

$$\therefore \quad RT\,\ell n(P^o/P) = \overline{V}(P' - P)$$

$$Rt\,\ell n(P^o/P) = \overline{V}\pi$$

For water $\overline{V} = 18$ cc

$$\therefore \quad \Delta G = Rt\,\ell n(P^o/P) = 18\text{cc} \times 2\text{ atm}$$

$$\Delta G = 36\text{ cc} - \text{atm.}$$

Q. 5 : At 25°C, sucrose and water form an ideal solution and the solubility of sucrose in the water is 6.2 molal. Calculate the heat of fusion of sucrose given that its molecular weight is 342, melting point is 200°C and the heat of fusion of water is 1400 cal/mole.

Solution: This problem can be considered a freezing point depression problem with sucrose as the solvent. The equation that relates the freezing point depression of a solution to the solubility is given by

$$-\ell n\,X_S = \frac{\Delta H_f}{R}(1/T - 1/T_f) \qquad (1)$$

where X_s is the mole fraction of sucrose. ΔH_f is the heat of fusion of sucrose. T_f is the melting point of sucrose. Solving for ΔH_f in equation (1), it becomes

$$\Delta H_f = -(\ell n\,X_S)\,(R)\,\frac{TT_f}{T_F - T} \qquad (2)$$

The mole fraction of a 6.2 molal solution can be calculated by the relation

$$X_S = \frac{n_S}{n_S + n_{H_2O}}$$

where n is the number of moles

$$X_S = \frac{6.2 \text{ moles/kg}}{(6.2 + 55.5) \text{ moles/kg}}$$

$$X_S = 0.10$$

From Equation (2)

$$\Delta H_f = (-\ell n\, 0.10 \text{ mol})\left(\frac{1.987 \text{ cal}}{\text{mol}\,^\circ\text{K}}\right)(298^\circ\text{K})\left(\frac{473}{175}\right)$$

or $\Delta H_f = 3650$ cal = 3.65 kcal.

Q. 6 : Calculate the freezing and boiling points of a solution containing 10 grams of $C_6H_{12}O_6$ and 40 grams of water.

Solution: In this problem, water is the solvent. When $C_6H_{12}O_6$ is added to the water, the resulting solution has a higher boiling point and a lower freezing point than that of the pure solvent. The difference between the boiling and freezing points of the solution and those of the pure solvent are the boiling point elevation, ΔT_B and the freezing point depression, ΔT_F respectively. ΔT_B and ΔT_F are directly proportional to the molality of the solution by the relations

$$\Delta T_B = K_B m$$

and

$$\Delta T_F = K_F m$$

where K_B and K_F are the molal boiling point constant and the molal freezing point constant respectively. m is the molality of the solute. For water

$K_B = 0.52$°C – kg H_2O/mole solute and

$$K_F = 1.86 \frac{^\circ\text{C} - \text{kg } H_2O}{\text{mole solute}}$$

But $\Delta T_B = K_B m$

$$= \left[\frac{0.52^\circ\text{C} - \text{kg } H_2O}{\text{moles } C_6H_{12}O_6}\right]\left[\frac{10.0\text{g } C_6H_{12}O_6}{40.0\text{g } H_2O}\right]$$

$$\left[\frac{10^3 \text{g } H_2O}{1\text{kg } H_2O}\right]\left[\frac{1 \text{ mole } C_6H_{12}O_6}{180\text{g } C_6H_{12}O_6}\right]$$

$\Delta T_B = 0.72$°C

$\therefore$ TB = 100°C + 0.72°C

$T_B = 100.72$°C.

$$\Delta T_F = K_F m$$

$$= \left[\frac{1.86°C - kg\ H_2O}{mole\ C_6H_{12}O_6}\right]\left[\frac{10.0g\ C_6H_{12}O_6}{40.0g\ H_2O}\right]$$

$$\left[\frac{10^3 g\ H_2O}{1kg\ H_2O}\right]\left[\frac{1\ mole\ C_6H_{12}O_6}{180g\ C_6H_{12}O_6}\right]$$

$$\Delta T_F = 2.58°C$$

$$\therefore \quad T_F = 0.00°C - 2.58°C$$

$$T_F = -2.58°C.$$

Q. 7 : Carbon dioxide obeys the reduced Berthelot equation reasonably well

$$z = \frac{PV}{RT} = 1 + \frac{9}{128}\frac{PT_C}{P_C T}\left[1 - 6\frac{T_C^2}{T^2}\right]$$

Given: $T_C = 304.3$ K and $P_C = 73.0$ atm, calculate the fugacity of carbon dioxide at a temperature of 150°C and pressure of 50 atm.

Solution: By definition, $d\,\ell n\, f = d\,\ell n\, P + \frac{1}{RT}(V - V_{id})\ dP$. On substituting the solution for V from the given equation and using the relation $V_{id} = \frac{RT}{P}$ Yields

$$d\,\ell n\, f = d\,\ell n\, P + \frac{9T_C}{128\ P_C T}\left[1 - 6\frac{T_C^2}{T^2}\right] dP \qquad (1)$$

where f = fugacity

P = pressure

T_C = critical temperature of temperature at the critical point

T = 150 C or, 423 K

Note that at a pressure approaching zero, the fugacity can be equated to the pressure, P.

Therefore, the integration of equation (1) gives

$$\ell n\, f = \ell n\, P + \frac{9T_C}{128\ P_C T}\left[1 - 6\frac{T_C^2}{T^2}\right] P$$

Substituting the numberical values

$$\ell n\,f = \ell n\,50 + \frac{(9)\,(304.3)}{(128)(73.0)(423)}\left[1 - \frac{6(304.3)^2}{(423)^2}\right]50$$

$$= 3.91\; +\; (0.0006929)\;(-2.105)\;50$$

and f = 46.4 atm.

Q. 8 : Determine the heat of vaporization of acetonitrile with a normal boiling point of 80C, given that acetonitrile has a vapor pressure which is changing at the rate of 0.030 atm/deg K around the boiling point.

Solution : The Clausius-Clapeyron equation relates the rate of changes of pressure and temperature to the heat of vaporization. Writing the equation in the form

$$\frac{dP}{dT} = \frac{P\Delta H_{vap}}{RT^2}, \qquad (1)$$

ΔH_{vap} can easily be calculated. The problem states that

$$\frac{dP}{dT} = 0.030\frac{atm}{deg\ K}$$

From Equation (1) therefore,

$$\Delta H_{vap} = \frac{dP}{dT}\,\frac{RT^2}{P_t}$$

At the boiling point P_t = 1 atm.

$$\therefore \Delta H_{vap} = \left[0.030\frac{atm}{^\circ K}\right]\frac{(1.987 cal\ mole^{-1}\ {}^\circ K^{-1})\ (353^\circ K)^2}{1\ atm}$$

$$= (7427.9 cal\ mol^{-1})\left[\frac{1\ kcal}{1000\ cal}\right]$$

$$= 7.4\ kcal\ mol$$

An alternative, less precise, method yields a similar result. The procedure used is Trouton's Rule which states that $\frac{\Delta H_{vap}}{T_b} = 21.$ cal mole^{-1} K^{-1}, where ΔH_{vap} is the heat of vaporization, and T_b is the normal boiling point temperature, in K. Hence,

$$\Delta H_{vap} = 21 cal\ mole^{-1}\ K^{-1}(353K)$$

$$= 7413\ cal\,/\,mole$$

$$= 7.4\ kcal\ mole^{-1}.$$

Q. 9 : By how much will 50 grams of water have its freezing point depressed if you add 30 grams (molecular weight 80) of glucose to it ?

Solution : The addition of any substance to water will alter its boiling or freezing point. To determine the amount of change, you must know the concentration of solute (the substance dissolved) in the solvent (water). This information is required because the freezing point depression, ΔT_f, equals the molal freezing point constant, k_f, times the molality (m)

$$\Delta T_f = k_f\,(m).$$

The concept of molality refers to the number of moles of solute per 1 Kilogram of solvent.

The solute, glucose, weighs 30 grams. Therefore, the number of moles of glucose is 30 g/180 g/mole = 1/6 moles. You have 50 grams of water. However, molality refers to moles per thousand grams. As such, a conversion is required; namely 50/1000. Hence, the molality is

$$\frac{\frac{1}{6}\text{ moles}}{\frac{50}{100}\text{g}} = 3.33\text{m}$$

Recall that the amount of depression of the freezing point is defined as $\Delta T_f = (k_f\ m)$. k_f is given for water as -1.86 deg mole^{-1}. You calculated m. The temperature depression is thus $\Delta T_f = k_f \times m = -1.86° \times 3.33\text{m} = -6.2°$.

Q. 10 : Find the vapor pressure of CCl_4 (carbon tetrachloride) at 38°C, if $H^1 = 32{,}000$ J/mole and at 23°C, the vapor pressure is .132 atm.

Solution: To solve this problem, employ the Clausius-Clapeyron equation, which permits evaluation of vapor pressure in terms of ΔH^1 (the heat required to transform one mole of liquid to the ideal-gas state), the original vapor pressure and the two temperatures. The equation states :

$$\log\frac{p_1}{p_2} = \frac{\Delta H^1}{19.15}\left[\frac{1}{T_2} - \frac{1}{T_1}\right],$$

where p_1 and p_2 are vapor pressures and T_1 and T_2 are the temperatures in Kelvin. Let p_2 be the vapor pressure of CCl_4 at 38°C.

Substitute the known values for solution,

$$\log\frac{.132}{p_2}=\frac{32,000}{19.15}\left[\frac{1}{311}-\frac{1}{296}\right].$$

Solve for p_2, to obtain $p_2 = .250$ atm.

Q. 11 : A sugar solution was prepared by dissolving 9.0 g of sugar in 500 g of water. At 27°C, the osmotic pressure was measured as 2.46 atm. Determine the molecular weight of the sugar.

Solution: The molecular weight of the sugar is found by determining the concentration, C, of sugar from the equation for osmotic pressure,

$$\pi = CRT,$$

where π is the osmotic pressure, R = universal gas constant = 0.08206 liter–atm/mole–°K, and T is the absolute temperature.

The osmotic pressure is measured as π = 2.46 atm and the absolute temperature is T = 27°C + 273 = 300°K, hence

$$\pi = CRT,$$

$$2.46 \text{ atm} = C \times 0.08206 \text{ liter-atm/mole-°K} \times 300\text{°K},$$

$$\text{or } C = \frac{2.46 \text{ atm}}{0.08206 \text{ liter - atm / mole-°K} \times 300\text{° K}}$$

$$= 0.10 \text{ mole / liter.}$$

It we assume that the volume occupied by the sugar molecules in so small a concentration can be neglected, then in 1 liter of solution there is approximately 1 liter of water, or 1000 g of water, and

$$C = 0.10 \text{ mole/1000g.}$$

Therefore, there is 0.10 mole of sugar dissolved in 1000g of water.

9.0g of sugar dissolved in 500g of water is equivalent to 18.0g of sugar dissolved in 1000g of water (9.0g/500g = 18.0g/1000g). But since C = 0.10 mole/1000g, the 18.0g of sugar must correspond to 0.10 mole of sugar. Therefore, the molecular weight of the sugar is

18g/0.1 mole = 180g/mole.

Q. 12 : What would be the surface tension of a liquid (density 0.876g/cm^3) which rises 2.0 cm in a capillary where water at 20°C rises 4.5 cm ?

Solution: Surface tension refers specifically to the force within a liquid that acts parallel to the surface and tends to stretch the surface out. The equation that relates surface tension to the density of a liquid is

$$\partial = 1/2\ \mathrm{rh\rho g}$$

where ∂ is the surface tension, r is the radius of the tube, h is the height the liquid rises, ρ is the density of the liquid and g is the downward acceleration of gravity.

This equation can be rewritten

$$\mathrm{rg} = \frac{2\partial}{\mathrm{h\rho}}$$

Because r and g are constant for a given capillary tube

$$\frac{2\partial_{H_2O}}{h_{H_2O}\ \rho_{H_2O}} = \frac{2\partial_{liq}}{h_{liq}\ \rho_{liq}}$$

One can find ∂H_2O and ρH_2O from standard tables.

∂H_2O = 72.62 dyn/cm, ρH_2O = 1 g/cm³.

Solving for liq:

$$\frac{2\ (72.62\mathrm{dyn/cm})}{(4.5\mathrm{cm})\left(1\mathrm{g/cm^3}\right)} = \frac{2\partial\ \mathrm{liq}}{(2.0\mathrm{cm})\left(0.876\mathrm{g/cm^3}\right)}$$

$$\partial\ \mathrm{liq} = \frac{2(72.62\ \mathrm{dyn/cm})(2.0\mathrm{cm})\left(0.876\mathrm{g/cm^3}\right)}{2(4.5\mathrm{cm})\left(1\ \mathrm{g/cm^3}\right)}$$

$$\partial\ \mathrm{liq} = 28.27\ \mathrm{dyn/cm}.$$

Q. 13 : Given that the surface tension of water is 72.62 dyn/cm at 20°C, how high should water rise in a capillary that is 1.0 mm in diameter ?

Solution: The surface tension is related to the radius of the capillary tube by the equation :

$$\gamma = \tfrac{1}{2}\ \mathrm{rh\rho g}$$

where γ is the surface tension, r is the radius of the capillary tube, h is the height the water rises, ρ is the density and g is the downward force of gravity. The density of H_2O is 1 g/cm³, the force of gravity is 980 cm/s² and the surface tension is given as 72.62 dyn/cm. The radius is found by dividing the diameter of the tube by 2. (1 mm = 10^{-1} cm). Solving for h:

$$72.62 \frac{\text{dyn}}{\text{cm}} = 72.62 \frac{\text{g}}{\text{sec}^2}$$

$$72.62 \text{ g/sec}^2 = ½ (10^{-1}/2 \text{ cm}) (\text{h}) (1 \text{ g/cm}^3) (980 \text{ cm/sec}^2)$$

$$\text{h} = \frac{72.62 \text{ g/sec}^2 (2)(2)}{(10^{-1}\text{cm})(1 \text{ g/cm}^3)(980 \text{ cm/sec}^2)}$$

$$\text{h} = 2.96 \text{ cm.}$$

Q. 14 : A wine has an acetic acid (CH_3COOH, 60 g/formula weight) content of 0.66% by weight. If the density of the wine is 1.11 g/ml, what is the formality of the acid ?

Solution: This problem involves the correct interpretation of percent by weight. 0.66% by weight means 0.66×10^{-2} g acid per 100g wine or 6.6×10^{-2} g acid per 1000g wine. To convert 6.6×10^{-2} g acid to formula weight we divide by 60g/formula weight and obtain 6.6×10^{-2} g/60 g/formula weight = 1.1×10^{-3} formula weight. To convert 1000g wine to volume, we divide by the density of the wine and obtain 1000 g/1.11 g/ml = 900 ml = 0.90 liter. The formality is then

$$\text{formality} = \frac{\text{formula weights}}{\text{volume (liters)}} = \frac{1.10 \times 10^{-3} \text{ formula weight}}{0.90 \text{ liter}}$$

$$= 1.2 \times 10^{-3} \text{ formal} = 1.2 \times 10^{-3} \text{F.}$$

Q. 15 : Calculate the molarity of a solution containing 10.0 grams of sulfuric acid in 500 ml of solution. (MW of H_2SO_4 = 98.1).

Solution: The molarity of a compound in a solution is defined as the number of moles of the compound in one liter of the solution. In this problem, one is told that there are 10.0 grams of H_2SO_4 present. One should first calculate the number of moles that 10.0 g represents. This can be done by dividing 10.0g by the molecular weight of H_2SO_4.

$$\text{number of moles} = \frac{\text{amount present in grams}}{\text{molecular weight}}$$

$$\text{number of moles of } H_2SO_4 = \frac{10.0\text{g}}{98.1\text{g/mole}} - 0.102 \text{ moles}$$

Since molarity is defined as the number of moles in one liter of solution, and since, one is told that there is 0.102 moles in 500 ml (½ of a liter), one should multiply the number of moles present by 2. This determines the number of moles in H_2SO_4 present in 1000 ml.

Number of moles in 1000 ml = 2 × 0.102 = 0.204.

Bucause molarity is defined as the number of moles in 1 liter, the molarity (M) here is 0.204 M.

Q. 16 : What is the molality of a solution in which 49g of H_2SO_4 (MW 98) is dissolved in 250 gram of water ?

Solution: Molality is defined as the number of moles of solute per kilogram of solvent.

$$\text{Molality} = \frac{\text{moles of solute}}{\text{no. of kg. of solvent}}$$

Here, the solute is H_2SO_4 and the solvent is water. In this problem, one is given the number of gram of solute and the number of grams of solvent. One must calculate the number of moles of solute and the number of kilograms of solvent.

The number of moles of solute (H_2SO_4) is found by dividing the number of grams available by the molecular weight.

$$\text{no. of moles} = \frac{\text{no. of grams}}{\text{MW}}$$

$$\text{no. of moles of } H_2SO_4 = \frac{49\text{g}}{98\text{g / mole}} = 0.5 \text{ moles}$$

Grams can be converted to kilograms by multiplying the number of grams by the conversion factor 1 kg/1000 gm. For the water, 250 × 1 kg/1000gm = .250kg. The molality now be found.

$$\text{Molality} = \frac{\text{no. of moles of } H_2SO_4}{\text{no. of kg. of } H_2O}$$

$$\text{Molality} = \frac{0.5 \text{ moles}}{.250 \text{ kg.}} = 2.0 \text{ moles / kg.}$$

Q. 17 : Determine the molarity of a 40.0% solution of HCl which has a density of 1.20 g/ml.

Solution: The molarity is defined as the number of moles of a compound present in 1 liter of solution. Here one is told that the density of this solution is 1.20 g/ml. This means that one ml of the solution weighs 1.20g. The solution is 40.0% HCl, thus 40.0% of the 1.20g is made up by HCl and 60% by H_2O. Here one finds the molarity of the HCl by: (1) determining the total weight of 1 liter fo the solution, (2) calculating the weight of HCl present, and (3) finding the number of moles of HCl present in 1 liter (molarity).

Solving: (1) If 1 ml of this solution weight 1.20g, 1000 ml (1 liter) is equal to 1000 × 1.20g.

weight of solution = density × 1000 ml

$= 1.20$ g/ml. × 1000 ml = 1200g

(2) 40.0% of this weight is taken up by HCl.

weight of HCl = weight of solution × .40

$= 1200g \times .40 = 480g.$

(3) The molecular weight of HCl is 36.5 g/mole, thus the number of moles of HCl present is this 1 liter of solution is equal to the weight of the HCl present divided by its molecular weight.

$$\text{no. of moles of HCL in 1 liter} = \frac{\text{weight of HCl}}{\text{MW}}$$

$$= \frac{480\text{g}}{36.5\text{g / mole}} = 12.15 \text{ moles}$$

The solution is therefore 13.15 moles/liter in HCl or 13.15M.

Q. 18 : You have 100-proof (50 percent alcohol by volume) bonded Scotch whisky. Calculate its molarity, mole fraction, and molality. If the temperature were to drop to – 10°C, could you still drink the Scotch? Assume density = .79 g/ml for ethyl alcohol (C_2H_5OH) and $K_f = 1.86$°C/m for water.

Solution: To find molarity, mole fraction, and molality, you need to know how many moles of alcohol are present in the whiskey. You can obtain this from a calculation using the density.

To find whether you can still drink the whisky at –10°C, requires you to know its freezing point, since you cannot drink something if it is frozen. You can determine the freezing point of the Scotch by determining the freezing point depression of the water containing the alcohol. This is found from $\Delta T_f = K_f\, m$, where ΔT_f = the depression, K_f = freezing point depression constant, and m = molality.

You now proceed as follows :

You are told that 100 proof means 50 percent alcohol by volume. This means that 1 liter of components possess 500 cm^3 of ethyl alcohol (a liter=1000 cm^3). Density=mass/volume. You know the density and volume, so that mass of alcohol = density × volume = .79 (500) = 395g.

Molecular wt. of the alcohol = 46.07g. Thus, you have 395/46.07 = 8.57 moles of alcohol. Hence,

$$\text{molarity} = \frac{\text{moles of solute}}{\text{liters of solution}} = \frac{8.57}{1} = 8.57\text{M}.$$

$$\text{Molality} = \frac{\text{moles of solute}}{\text{kilograms of solvent}}.$$

To find the number of grams of solvent, you need to use the density again. In 1 liter of Scotch Whisky, you have 500 cm^3 of water. Density of water = 1 g/cm^3.

Thus, mass of water = 500 cm^3 (1 g/cm^3) = 500 grams. Thus,

$$\text{Molality} = \frac{8.57}{\frac{500}{1000}} = 17.14\text{m}$$

you divided 500 by 1000 since molatity is moles per 1000g of solvent.

Mole fraction: (MW of H_2O=18.) moles of H_2O 500g/18.0 g/mole.

Mole fraction of ethyl alcohol =

$$\frac{\text{moles ethyl alcohol}}{\text{moles ethyl alcohol} + \text{moles } H_2O} = \frac{8.57}{8.57 \times \frac{500}{18}} = .2357.$$

Now, you answer whether the scotch can be consumed at 10°C. You already calculated the molality. Thus, $\Delta T_f = k_f m$ = (1.86) (17.14) = 31.88°C. Thus, the freezing point of whisky is 0°C – 31.88°C –31.88°C. The temperature is only –10°C, however. This means whisky is still a liquid and can be consumed.

Q. 19 : If 20 ml of 0.5 N salt solution is diluted to 1 liter, what is the new concentration ?

Solution: When considering normality one must always keep in mind that it is a concentration defined as the number of equivalents per liter. Since the number of equivalents does not change during dilution, equivalents before dilution = equivalents after dilution, or in other words,

$$N_1V_1 = N_2V_2 \qquad 1 \text{ liter} = 1000 \text{ ml}$$

where N_1 is the normality of the initial solution, V_1 the initial volume, N_2 the final normality and V_2 is final volume.

Solving for the final normality :

$$(0, 5)\ (0.020\ell) = N_2\ (1\ell)$$

$$N_2 = \frac{(0.5)\ (0.020\ell)}{1\ \ell} = 0.01N$$

Q. 20 : Calculate the volume of 0.3 N base necassary to neutralize 3 liters of 0.01 N nitric acid.

Solution: For neutralization to occur, there must be the same number of hydrogen ions as there are hydroxide ions. This is shown by the following equation

$$H^+ + OH^- \leftrightarrows H_2O$$

The H^+ ions come from the acid, the OH^- ions from the base. The number of H^+ and OH^- ions are equal to the number of equivalents.

Normality is defined as the number of equivalents of acid or base per liter of solution.

$$\text{Normality} = \frac{\text{equivalents}}{\text{liter}}$$

An equivalent is the number of grams of the acid or base multiplied by the number of replaceable hydrogens or hydroxides divided by the molecular weight of the acid or base.

$$\text{equivalent} = \frac{\text{grams of solute} \times \text{no. of replaceable H or OH}}{\text{MW of solute}}$$

The number of replaceable hydrogens or hydrodixes is defined as the number which ionize when the compound is placed in solution. For nitric acid, there is one replaceable hydrogen

$$HNO_3 \leftrightarrows H^+ + NO_3^-$$

In brief, for neutralization there must be the same equivalent amount of base as there is acid. Because normality is defined as

$$\text{Normality} = \frac{\text{equivalents}}{\text{liters}}$$

equivalents are equal to the normality times the volume.

Thus, the normality of the base times its volume equals the normality of the acid times its volume.

$$N_{base}\ V_{base} = N_{acid}\ V_{acid}$$

where N is the normality and V is the volume.

Here, $0.3 \text{ equiv / liter} \times V_{base} = 0.01 \text{ equive / liter} \times 3 \text{ liters}$

$$V_{base} = \frac{0.01 \times 3 \text{ liters}}{0.3}$$

$$= 0.1 \text{ liter of } 100 \text{ M.}$$

Q. 21 : Given an aqueous solution of NH_3 at 70°F (22.11°C), the partical pressure of water over the solution is 0.34 psi at 0.05 mole fraction. The partial pressure of HN_3 over the same solution is 0.83 psi. Calculate the activity coefficient of water, NH_3 and the activity of water and NH_3. The vapor pressure of pure water at 70°F is 18.77 torr, and Henry's constant, K_2 for NH_3 is 725 torr.

Solution:

Water : Activity coefficient, γ_i is defined as

$$\gamma_i = \frac{P_i}{x_i P_i^o}$$

where P_i = partial pressure of component i

P_i^o = vapor pressure of component i

x_i = mole fraction of component i

$$\therefore \quad \gamma_{H_2O} = \frac{(0.34\,\text{pis})\left(\frac{1\text{ atm}}{14.6960\text{ pis}}\right)\left(\frac{760\text{ torr}}{1\text{ atm}}\right)}{(0.95)\,(18.77\text{ torr})}$$

$$= 0.986$$

The activity, a_i of a nonelectrolyte is equal to the product of the activity coefficient and the mole fraction.

Thus $\quad a_i = \gamma_i\, x_i$

$$\therefore \quad a_{H_2O} = (0.986)\,(0.95)$$

$$= 0.937$$

NH_3: Activity coefficient of NH_3 using Henry's law constant is defined as

$$\gamma_i = \frac{P_i}{K_i x_i}$$

where K_i is Henry's low constant

$$\therefore \quad \gamma_{NH_3} = \frac{(0.83\,\text{psi})\left(\frac{1\text{ atm}}{14.6960\text{ psi}}\right)\left(\frac{760\text{ torr}}{1\text{ atm}}\right)}{(0.05)\,(725\text{ torr})}$$

$$= 1.18$$

$$a_{NH_3} = \gamma_{NH_3} x_{NH_3} = (1.18)\,(0.05) = 0.059$$

Note that the subscript 2 in K_2 refers to the Henry constant of the solute, in this case, NH_3.

Q. 22 : Henry's Law constant for hydrogen gas in water is 5.34×10^7 torr and 2.75×10^6 torr in benzene. How much more soluble is hydrogen in benzene than in water ?

Solution: Henry's Law constant K_2 is given by

$$K_2 = \frac{P_2}{x_2}$$ (valid in very dilute solution in the region where $x_2 \to 0$)

otherwise written as $K_2 = \lim_{x_2 \to 0} \frac{P_2}{x_2}$

where P_2 = Partial pressure of hydrogen gas over the solution
x_2 = concentration or solubility

The subscript 2 indicates that the solute which is the component at lower concentration is being considered.

$$\therefore \quad K_{2(C_6H_6)}X_{2(C_6H_6)} = P_{2(C_6H_6)}$$

and $$K_{2(H_2O)}X_{2(H_2O)} = P_{2(H_2O)}$$

$$\therefore \quad K_{2(C_6H_6)}X_{2(C_6H_6)} = K_{2(H_2O)}X_{2(H_2O)} \qquad (1)$$

Taking the ratio of x_2 in the solvents as given by (1),

$$\frac{x_{2(C_6H_6)}}{x_{2(H_2O)}} = \frac{K_{2(H_2O)}}{K_{2(C_6H_6)}} = \frac{5.34 \times 10^7}{2.75 \times 10^6} = 19.4$$

H_2 is thus 19.4 times more soluble in benzene than in water.

Q. 23 : Iodine is added to a solution of water and carbon disulfied. The distribution coefficient, K_d, for I_2 between water and carbon disulfied is 410. After an amount of CS_2 equal to the water equilibrates with the aqueous phase, calculate the fraction of iodine remaining in the water phase.

Solution: Since the amount of substance of solute added is small, the distribution coefficient, K_d, is given by the ratio of the concentrations of the solute in the two phases.

$$\therefore \quad K_d = \frac{C'}{C} \qquad (1)$$

where C' = concentration of I_2 in the CS_2 layer
and C = concentration of I_2 in the water layer

The fraction of I_2 in the two phases must sum up to unity, i.e.

$$n_2' + n_2 = 1 \qquad (2)$$

where n_2' = fraction of I_2 in the CS_2 layer
n_2 = fraction of I_2 in the water layer

$$\therefore \quad n_2' = \frac{n_2'}{V'}$$

and $$C_2 = \frac{n_2}{v}$$

where v' and v are the volume CS_2 and water respectively.

For (1) $$K_d = \frac{C_2'}{C_2} = \frac{n_2'/v'}{n_2/v} = 410$$

But $$v' = v \quad \therefore \quad \frac{n_2'}{n_2} = 410 \text{ and } n_2' = 410n_2$$

From (2) $$n_2' + n_2 = 1$$

$$\therefore \quad 410n_2 + n_2 = 1$$

$$411n_2 = 1$$

or $$n_2 = \frac{1}{411} = 2.43 \times 10^{-3}$$

i.e., 2.43×10^{-3} of the original amount of I_2.

Q. 24 : The partial molar volume V_2 of K_2SO_4 in water solutions at 298 K is given by

$$\overline{V}_2(cm^3) = 32.280 + 18.216m^{½} + 0.0222m$$

Obtain an equation for $\overline{V}_1$ the partical molar volume of H_2O. Take $\overline{V}_i$ = 17.963 cm^3. mol^{-2} for H_2O.

Solution: In a binary system, the partial molar volumes cannot vary independently of one another. If, for some reason, $\overline{V}_1$ increases, then $\overline{V}_2$ must decrease. This can be represented mathematically by

$$n_1 d\overline{V}_1 = -n_2 d\overline{V}_2 \qquad (2)$$

where n_1 = number of moles of component 1
and n_2 = number of moles of component 2.

Equation (1) is known as the Gibbs-Duhem equation. $\overline{V}_2$ is given to be 32.280 + 18.216$m^{½}$ + 0.0222m. Therefore

$$d\overline{V}_2 = \frac{1}{2}\left(18.216m^{-½}\right) dm + 0.0222dm$$

$$= \left(9.108m^{-½} + 0.0222\right)dm$$

$$\text{where m} = \text{molality} = \frac{\text{g of solute}}{\text{1 kg of solvent}}.$$

Replace n_2 by m and n_1 by $\frac{1000}{M_1}$ where M_1 = molecular weight of component 1 which is water. From equation (1)

$$d\overline{V}_1 = -\frac{n_2}{n_1} d\overline{V}_2 = -\frac{mM_1}{1000}\left(9.108m^{-½} + 0.0222\right)dm$$

$$= -\frac{M_1}{1000}\left(9.108m^{½} + 0.0222m\right)dm$$

Integration both sides with respect to m gives

$$\overline{\ }_1 = -\frac{M_1}{1000}\int\left(9.108m^{½} + 0.0222m\right)dm. + C \qquad (2)$$

where C is an integration constant.

$$\therefore\ \overline{V}_1 = -\frac{M_1}{1000}\left(6.072m^{3/2} + 0.0111m^2\right) + C. \qquad (3)$$

Using the fact that $\overline{V}_1 = \overline{V}_i$ = initial partial molar volume of water (that is when m = 0), equation (3) becomes

$$17.963 = 0 + C$$

$$\therefore \qquad C = 17.963.$$

Since M_1 = molecular weight of H_2O = 18, equation (3) gives

$$\overline{V}_1 = -0.1094m^{3/2} - 0.0002m^2 + 17.963$$

which is the desired equation.

7

Chemical Equilibrium

Q. 1 : $\Delta G°$ is 300 cal at 40°C for the reaction :

$$2NO_2 = N_2O_4$$

(a) Calculate K_p for this reaction at 40°C.

(b) The density of an equilibrium mixture of NO_2 and N_2O_4 gases is found to be 5.85 g/liter at 40°C and a certain pressure. Using the value in (a) for K_p, calculate the degree of dissociation of the N_2O_4, the average molecular weight of the mixture, and the total pressure of the mixture.

Solution: (a) Looking at the data given in the problem, the best formula for calculating K_p is $\Delta G° = -RT \ln K_p$ or

$$\ln K_p = -\frac{\Delta G°}{RT} \quad (1)$$

where $\Delta G°$ = free energy change, R = gas constant and T = temperature.

Here, $\Delta G° = 300$ cal

$R = 1.98$ cal mole^{-1} degree^{-1}

$T = 40°C = (40 + 273)°K = 313°K$

Substituting these values into equation (1), gives

$$\ln K_p = -\frac{300}{(1.98)\,(313)} = -0.484$$

$$\therefore \qquad K_p = e^{-0.484} = 0.62$$

$$\text{or } 2.303 \log K_p = -\frac{300}{(1.98)(313)}$$

$$\log K_p = -\frac{300}{(1.98)(313)(2.303)} = -0.21$$

$$\text{or} \qquad K_p = 0.62.$$

(b) From the ideal gas law,

$$PM = \rho RT \tag{2}$$

where P = Pressure

M = Average molecular weight

ρ = Density = 5.85 g/liter

R = Gas constant = 0.082

T = Temperature = 313

Therefore $PM = (5.85)(0.082)(313) = 150$g atm/mole. Let n° denote the number of moles if the N_2O_4 were completely undissociated and α, the degree of dissociation. Then $n_{N_2O_4} = (1-\alpha)n^\circ$, $n_{NO_2} = 2\alpha n^\circ$ and $N = (1+\alpha)n^\circ$. The average molecular weight M is :

$$M = 46N_{NO_2} + 92N_{N_2O_4} \text{ where } N_{NO_2} = \text{mole fraction of } NO_2 = \frac{2\alpha n^\circ}{(1+\alpha)n^\circ}$$

$$\text{or} \quad N_{N_2O_4} = \text{mole fraction of } N_2O_2 = \frac{(1-\alpha)n^\circ}{(1+\alpha)n^\circ}$$

$M = 46\dfrac{2\alpha}{1+\alpha} + 92\dfrac{1-\alpha}{1+\alpha} = 92/(1+\alpha)$. Since $P = 150/M$, we find $P = (1+\alpha)(150/92)$. Finally, the equlibrium constant is

$$K_P = \frac{n_{N_2O_4}}{n^2_{NO_2}}\frac{N}{P} = (1-\alpha)(1+\alpha)/4\alpha^2 P$$

$$\text{or} \qquad 0.62 = 92\,(1-\alpha)/150 \times 4\alpha^2$$

Then $(1-\alpha)/\alpha^2 = 4.04$. Use the formula $\alpha = \dfrac{-b \pm \sqrt{b^2 - 4ac}}{2a}$

to solve for α

$$\therefore \qquad \alpha = \frac{-1 \pm \sqrt{1-16.16}}{8.08}$$

Picking the positive value for α given α = 0.39. M is then 92/1.39 = 66, and P = 150/66 = 2.27 atm.

Q. 2 : K_p has the value 10^{-5} for the equilibrium

$$CO_2\,(g) + H_2\,(g) + CO(g) + H_2O(g)$$

at 25°C, and ΔS°, the entropy change, is –10 cal/deg mole for the reaction as written (ΔH° and ΔS° do not change much with temperature). One mole of CO, 2 moles of H_2, and 3 moles of CO_2 are introduced into a 5-liter flask at 25°C. Calculate (a) Δ 90°, the free energy change at 25°C, (b) the equlibrium pressure, (c) the moles of each species present at equilibrium, and (d) K_p at 100°C.

Solution: There are various expressions for K_p, one of which is

$$\text{In } K_p = -\frac{\Delta G^\circ}{RT} \qquad (1)$$

$$\therefore \qquad \Delta G^\circ = -RT \text{ In } K_p$$

$$= -(1.98)(298)(\text{In } 10^{-5})$$

$$= 6.78 \text{ kcal/mole for the reaction as written.}$$

(b) Using PV = nRT, we have P = nRT/V. Since there is no change in the number of moles, there will still be 6 moles of gas. Therefore, n = 6. Now

$$P = \frac{(6)\,(0.082)(298)}{5} = 29.323 \text{ atm}$$

(c) Let x = number of moles of H_2O

What is present at equilibrium from what is introduced initially can be represented by

Initially	3 moles		2 moles		1 mole		0 moles
	$CO_2(g)$	+	$H_2(g)$	=	CO(g)	+	$H_2O(g)$
at equilibrium	(3–x) moles		(2–x) moles		(1+x) moles		(x) moles

This change comes about from the stoichiometry of the reaction where each mole of CO_2 and each mole of H_2 consumed produces a mole of CO and a mole of H_2O. Now the K_p expression is represented by: $K_P = \frac{P_{H_2O}\,P_{CO}}{P_{CO_2}\,P_{H_2}}$ where P_i is the partial pressure of the respective gases. P_i = mole fraction of i × P_{tot} where P_{tot} is the total pressure = 29.3 atm.

The mole fraction of $i = \frac{\text{moles of i}}{\text{total number of moles}}$,

where total moles = 6 moles. Hence,

$$K_P = \frac{(x/6)\,P_{tot}\left(\frac{1+x}{6}\right)P_{tot}}{\left(\frac{3-x}{6}\right)P_{tot}\left(\frac{2-x}{6}\right)P_{tot}} = 10^{-5}$$

$$10^{-5} = \frac{x(1+x)}{(3-x)\,(2-x)} = \frac{x+x^2}{6-5x+x^2}$$

$$\left(x^2-5x+6\right)10^{-5} = x+x^2$$

$$\frac{x^2}{10^5} - \frac{5x}{10^5} + \frac{6}{10^5} - x - x^2 = 0$$

Approximately, $x = 6 \times 10^{-5}$. Therefore moles of CO_2, H_2, CO and H_2O are 3, 2, 1 and 6×10^{-5} respectively.

(d) $\Delta H° = \Delta G° + T\Delta S°$ (2)

Substituting values at 25°C = 298°K into equation (2), we have $\Delta H° = 6.78 + (298)\,(-10/1000)$.

$\Delta H° = 3.80$ kcal/mole.

Since the problem requires K_p at 100°C = 373°K, we first compute $\Delta G°$ at 373°K and the substitute its value into equation (1).

$$\Delta G°_{373} = \Delta H° - T\Delta S°$$
$$= 3.80 - (373)\,(-10/1000)$$
$$= 7.53 \text{ kcal / mole}$$

Thus $\ln K_P = \frac{-\Delta G^\circ}{RT} = -\frac{7.53 \text{ kcal / mole}}{(373^\circ K) \times 1.987 \times 10^{-3} \text{ kcal / mole } ^\circ K}$

$\therefore \quad K_P = 3.86 \times 10^{-5}$.

Q. 3 : In a study of the equilibrium, $H_2 + I_2 = 2HI$, a certain number of moles x of HI are formed when 1 mole of H_2 and 3 moles of I_2 are introduced into a flask of volume V at temperature T. On introducing 2 additional moles of H_2, the amount of HI formed is found to be 2x. Calculate K_p.

Solution: Let n = number of moles at equilibrium, then

$$n_{HI} = x$$

$$n_{H_2} = 1 - x/2$$

and $\quad n_{I_2} = 3 - x/2$

Now, if $\quad n_{HI} = 2x$,

$$n_{H_2} = 2 + (1 - 2x/2) = 3 - x$$

and $\quad n_{I_2} = 3 - 2x/2 = 3 - x$.

Since there are an equal number of moles of gaseous species on both sides of the equation $K_p = K_n$, hence

$$\frac{x^2}{(1 - x/2)(3 - x/2)} = \frac{(2x)^2}{(3 - x)^2}$$

Solving for x yields

$$x^2(3 - x)^2 = (2x)^2(1 - x/2)(3 - x/2)$$

$$x^2(9 - 6x + x^2) = 4x^2(3 - 2x + x^2/4)$$

$$9 - 6x + x^2 = 4(3 - 2x + x^2/4)$$

$$\therefore \quad \frac{9 - 6x + x^2}{3 - 2x + x^2/4} = 4$$

$$9 - 6x + x^2 = 12 = 8x + x^2$$

$$9 - 6x + x^2 - 12 + 8x - x^2 = 0$$

$$-3 + 2x = 0$$

or $x = 1.5$. Now,

$$K_P = \frac{(1.5)^2}{(1-1.5/2)\,(3-1.5/2)} = \frac{2.25}{(.25)\,(2.25)} = \frac{1}{.25} = 4.$$

Q. 4 : A 2-liter flask maintained at 700°K contains 0.1 mole of CO and a catalyst for the reaction

$$CO\,(g) + 2H_2(g) = CH_3OH(g)$$

Hydrogen is introduced until the equilibrium total pressure is 7 atm, at which point 0.06 moles of methanol are formed.

(a) Calculate K_p in atm^{-2}

(b) What would the final pressure be were the same amounts of CO and H_2 used but with no catalyst present so that no reaction occurs ?

Solution: Using the equation $PV = nRT$

where
P = total pressure
V = volume
n = final total moles
R = gas constant
T = temperature

We can solve for n, the final total moles. This value will enable us to compute the individual moles and finally K_P.

$$n = \frac{PV}{RT} = \frac{(7)\,(2)}{(0.082)\,(700)} = 0.244 \text{ moles.}$$

In this problem, 0.06 moles of methanol are formed. So the number of moles of CO and H_2 are 0.244 – 0.06 = 0.184. However, the number of moles of the CO must be 0.1 – 0.06 = 0.04, so number of moles of H_2 must be 0.184 – 0.04 = 0.144. Therefore, the equilibrium constant

$$K_P = \frac{(0.06)\,(0.244)^2}{(0.04)\,(0.144)^2(7)^2} = 0.088 \text{ atm}^{-2}$$

(b) Had no reaction occurred, the moles of CO would ramin

0.1, and the moles of H_2 would be 0.144 plus twice the moles of methanol in (a), or 0.144 + 0.12 = 0.264. The total moles are then 0.364 and

$$P = \frac{nRT}{V} = \frac{(0.364)\,(0.082)\,(700)}{2}$$

$$= 10.45 \text{ atm.}$$

Q. 5 : At a certain temperature, an equilibrium mixture of

$$NO_2 + SO_2 \rightleftarrows NO + SO_3$$

is analyzed and found to contain the following molar concentrations :

$$[NO_2] = 0.100,\ [SO_2] = 0.300,\ [NO] = 2.00,\ [SO_3] = 0.600.$$

If 0.500 moles of SO_2 are introduced at constant temperature, what will be the new concentrations of reactants and products when equilibrium is re-established ?

Solution: Le Chatelier's principle states: If a stress is placed on a system in equilibrium, whereby the equilibrium is altered, that change will take place which tends to relieve or neutralize the effect of the added stress. Thus, in this reaction, if more SO_2 is added more NO and SO_3 will be formed. If stress is placed on the left side of the equation, the reaction will be forced to the right (and vice versa). One can determine the equilibrium constant (Keq) for this reaction by using the concentrations of the original mixture. The equilibrium constant is defined :

$$Keq = \frac{[NO]\,[SO_3]}{[NO_2]\,[SO_2]}$$

where [ ] indicate concentrations. Solving for Keq with

$$[NO] = 2.00 \qquad [NO_2] = 0.100$$

$$[SO_3] = 0.600 \qquad [SO_2] = 0.1300,$$

One obtains

$$keq = \frac{(2.00)\,(0.699)}{(0.100)\,(0.300)} = 40.0.$$

One can solve for the new concentrations by using the Keq. From Le Chatelier's principle, one knows that when SO_2 is added to this mixture, the amounts of NO and SO_3 will increase. For each mole of SO_3 and NO formed, one mole of SO_2 and one mole will increase. For each mole of SO_3 and NO formed, one mole of SO_2 and one mole of NO_2 will react, thus the new concentrations of these components will be qual to the original concentrations less x moles. The new concentrations can be stated.

$$[NO] = 2.00 + x$$
$$[SO_3] = 0.600 + x$$
$$[NO_2] = 0.100 - x$$
$$[SO_2] = 0.300 + \text{the amount added} - x$$
$$= 0.300 + 0.500 - x = 0.800 - x.$$

Using the formula for the equilibrium constant, one can solve for x.

$$Keq = \frac{[NO][SO_3]}{[NO_2][SO_2]} = 40.0.$$

Substituting,

$$40.0 = \frac{(2.00 + x)(0.600 + x)}{(0.100 - x)(0.800 - x)}$$

$$40.0 = \frac{1.20 + 2.6x + x^2}{.80 - 0.90x + x^2}$$

$$(0.80 - 0.90x + x^2)40 = 1.20 + 2.6x + x^2$$

$$3.20 - 36.0x + 40x^2 = 1.20 + 2.6x + x^2$$

$$2.0 - 38.6x + 38x^2 = 0$$

One can use the quadratic formula to solve for x.

$$ax^2 + bx + c = 0$$

$$x = \frac{-b \pm \sqrt{b^2 - 4ac}}{2a}$$

$$38x^2 - 38.6x + 2.0 = 0$$

$$x = \frac{38.6 \pm \sqrt{(38.6)^2 - 4 \times 2.0 \times 38}}{2 \times 38}$$

$$x = \frac{38.6 \pm 34.44}{76}$$

$$x = \frac{38.6 + 34.44}{76} = 0.96$$

or $$x = \frac{38.6 - 34.44}{76} = 0.055$$

One cannot use x = 0.96 because $[NO_2]$ and $[SO_2]$ will be negative.

Concentration cannot have negative values, which means x = .05. One can now find the new concentrations

$$[NO] = 2.00 + x = 2.055 \text{ moles}$$

$$[SO_3] = 0.600 + x = 0.65 \text{ moles}$$

$$[NO_2] = 0.100 - x = 0.045 \text{ moles}$$

$$[SO_2] = 0.800 - x = 0.745 \text{ moles}$$

Q. 6 : Two moles of gaseous NH_3 are introduced into a 1.0-liter vessel and allowed to undergo partial decomposition at high temperature according to the reaction

$$2NH_3(g) \rightleftarrows N_2(g) + 3H_2(g).$$

At equilibrium, 1.0 mole of NH_3 (g) remains. What is the value of the equilibrium constant ?

Solution: This problem involves substitution into the expression for the equilibrium constant for this reaction,

$$K = \frac{[N_2][H_2]^3}{[NH_3]^2}$$

Since 1.0 moles of H_3 remains, 2.0 – 1.0 = 1.0 mole of NH_3 was consumed. Also, since one mole of N_2 and three moles of H_2 are formed per 2 moles HN_3 consumed, at equilibrium, there are 3/2 moles of H_2 and 1/2 mole of N_2. The equilibrium concentrations are therefore $[NH_3]$ = 1.0 mole/1.0 liter = 1.0m, $[N_2]$ = (1/2 mole)/(1.0 liter) = 0.5m, and $[H_2]$ = (3/2 moles)/(1.0 liter) = 1.5M.

Substituting these into the expression for K gives

$$K = \frac{[N_2]\,[H_2]^3}{[NH_3]^2} = \frac{(0.5)\,(1.5)^3}{(1)^2}$$

$$= 1.6875.$$

Q. 7 : Four moles of PCl_5 are placed in a 2-liter flask. When the following equilibrium is established

$$PCl_5 \rightleftarrows PCl_3 + Cl_2,$$

the flask is found to contain 0.8 mole of Cl_2. What is the equilibrium constant ?

Solution: The equilibrium constant (Keq) for this reaction is:

$$Keq = \frac{[PCl_3]\,[Cl_2]}{[PCl_5]}$$

where [ ] indicate concentrations. One is given that, at equilibrium, there are 0.8 mole of Cl_2 present. From the equation for the reaction, it follows that there is 0.8 mole PCl_3 present also. This means that 0.8 moles of PCl_5 has reacted. Originally 4 moles of PCl_5 were present, thus, at equilibrium, 4.0 – 0.8 moles or 3.2 moles are present. These components are present in a two-liter flask. Therefore, the concentrations will be expressed in moles/2 liters. One can now solve for the equilibrium constant.

$$Keq = \frac{[PCl_3]\,[Cl_2]}{[PCl_5]} \qquad \begin{array}{c} [PCl_3] = 0.8 \text{ moles / 2liters} \\ [Cl_2] = 0.8 \text{ mole / 2liters} \\ [PCl_5] = 3.2 \text{ moles / 2liters} \end{array}$$

$$Keq = \frac{(0.8/2)\,(0.8/2)}{(3.2/2)} = \frac{(0.4M)\,(0.4M)}{(1.6M)}$$

$$\frac{0.16M^2}{1.6M} = 0.1M.$$

Q. 8 : For the formation of hydrogen iodide,

$$H_2(g) + I_2(g) \rightleftarrows 2HI(g),$$

the value of the equilibrium constant at 700K is 54.7. If one mole of H_2 and one mole of I_2 are the only materials initially present, what will be the equilibrium concentrations of H_2, I_2, and HI.

Solution: This problem is an application of the expression for the equilibrium constant

$$k = \frac{[HI]^2}{[H_2][I_2]},$$

except that the concentrations are unknown. However, as we shall see, we do not need to know the concentrations.

Let x be the number of moles of H_2 that disappeared at equilibrium. Since one mole of I_2 is consumed and two moles of HI are formed per mole of H_2 consumed, then, at equilibrium, x moles of I_2 will have reacted, 2x moles of HI will have formed, and 1–x moles of H_2 and 1–x moles of I_2 will remain. If we let the volume of the vessel in which the reaction takes place be v, then $[H_2] = (1-x)/v$, $[I_2] = (1-x)/v$, and $[HI] = 2x/v$. The expression for the equilibrium constant is then

$$K = 54.7 = \frac{[HI]^2}{[H_2]\,[I_2]} = \frac{(2x/v)^2}{[(1-x)/v]\,[(1-x)/v]} = \frac{4x^2}{(1-x)^2}\,\frac{v^2}{v^2}$$

$$= \frac{4x^2}{1-2x+x^2},$$

or $$54.7 - 2\,(54.7)\,x + 54.7x^2 = 4x^2$$

$$50.7x^2 - 109.4x + 54.7 = 0.$$

Using the quadratic formula,

$$x = \frac{109.4 \pm \sqrt{(-109.4)^2 - 4(50.7)\,(54.7)}}{2\,(50.7)}$$

or $$x = 0.79,\ x = 1.37.$$

Since x, the number of moles of H_2 or I_2 that reacted, cannot be greater than one, the answer x = 1.37 must be a nonphysical entity, and we retain only the answer x = 0.79. Hence, at equilibrium,

$$[H_2] = 1 - x = 1 - 0.79 = 0.21 \text{ mole}$$

$$[I_2] = 1 - x = 1 - 0.79 = 0.21 \text{ mole}$$

$$[HI] = 2x = 2(0.79) = 1.158 \text{ moles.}$$

Q. 9 : Given that k = 8.85 at 298°K and k = .0792 at 373°K, calculate the ΔH° for the reaction of the dimerization of NO_2 to N_2O_4. Namely, $2NO_2\ (g) \rightleftarrows N_2O_4\ (g)$.

Solution: ΔH° is the standard enthalpy change, a measure of the heat content of the system. It is quantitatively related to the equilibrium constants of a system at different temperations by the van't Hoff equation :

$$\log\frac{k_2}{k_1} = \frac{\Delta H°}{19.15}\left[\frac{1}{T_1} - \frac{1}{T_2}\right],$$

where k_2 and k_1 are equilibrium constants, and T_1 and T_2 are temperatures in Kelvin. given that k_1 = 8.85 at T_1 = 298°K and k_2 = 0.0792 at 373°K, one can find ΔH° by the substitution of these values in the equation. Thus,

$$\log\frac{.0792}{8.85} = \frac{\Delta H°}{19.15}\left[\frac{1}{298} - \frac{1}{373}\right].$$

Solving for ΔH°, one obtains

$$\Delta H° = -\ 58{,}200 \text{ J/mole for the reaction.}$$

Q. 10 : Ammonia is synthesized commercially according to the equation

$$\tfrac{1}{2}\, N_2\ (g) + 1\tfrac{1}{2}\, H_2\ (g) \rightarrow NH_3\ (g),$$

under a constant total pressure of 50 atm. At 350°C the equilibrium constant has been determined as 0.0278. The heat of reaction for this process is ΔH = –11.0 kcal/mole. Determine the value of equilibrium constant at 450°C.

Solution: For a reaction with equilibrium constant k_1 at absolute temperature T_1 and equilibrium constant k_2 at temperature T_2,

$$\log\frac{k_2}{k_1} = \frac{-\Delta H°}{2.303R}\left[\frac{1}{T_2} - \frac{1}{T_1}\right],$$

where R is the gas constant.

Let $T_1 = 350°C = 623°K$, $k_1 = 0.0278$, and $T_2 = 450°C = 723°K$. $\Delta H° = -11.04$ Kcal/mole $= -11{,}040$ cal/mole.

Then

$$\log\frac{k_2}{k_1} = \frac{-\Delta H°}{2.303R}\left[\frac{1}{T_2} - \frac{1}{T_1}\right]$$

$$\log\left[\frac{k_2}{0.0278}\right] = \frac{11{,}040 \text{ cal / mole}}{2.303 \times 1.987 \text{ cal / mole - deg}}\left[\frac{1}{723° \text{ K}} - \frac{1}{623° \text{ K}}\right]$$

$$= -0.535.$$

Solving for k_2,

$$\log\left[\frac{k_2}{0.0278}\right] = -0.535$$

$$\frac{k_2}{0.0278} = 10^{-0.535} = 0.292$$

$$k_2 = 0.0278 \times 0.292 = 0.00812.$$

Thus, the value of the equilibrium constant at 450°C. is 0.00812.

Q. 11 : Calculate $\Delta G°$ for the reaction

$$N_2\,(g) + 3H_2\,(g) = 2NH_3\,(g)$$

at 400°C = 673°K using K_C. The value of K_P for this reaction is 1.64×10^{-4}. Interpret the different values of $\Delta G°$ calculated from K_P and K_C.

Solution: Given the equilibrium reaction :

(i) $aA_{(g)} + bB_{(g)} + \ldots = cC_{(g)} + dD_{(g)} + \ldots$

there are two constants that measure the equilibrium: K_C relates the concentration.

(ii) $$K_C = \frac{[C]^c[D]^d \ldots}{[A]^a[B]^b \ldots}$$

K_P relates the pressures

(iii) $$K_P = \frac{P_C^c\, P_D^d \ldots}{P_A^a\, P_B^b \ldots}$$

To derive K_C from K_p substiute for each pressure P = cRT, where R is the gas constant; T = absolute temperature and c is the number of moles of a gas per liter.

For the reaction,

$$N_2\,(g) + 3H_2(g) = 2NH_3\,(g),$$

$$K_p = \frac{P^2_{NH_3}}{P_{N_2}P^3_{H_2}} = \frac{C^2_{NH_3}(RT)^2}{C_{N_2}C^3_{H_2}(RT)^4} = \frac{K_C}{(RT)^2},$$

$$\text{where}\quad K_C = \frac{P^2_{NH_3}}{P_{N_2}P^3_{H_2}} = (RT)^2 K_P.\ \text{Thus,}$$

$$K_C = \left[\left[0.082\frac{\text{liter}-\text{atm}}{{}^\circ\text{K}-\text{mole}}\right](673^\circ\text{K})\right]^2 (1.64\times10^{-4}) = 0.500$$

$$\begin{aligned}\text{and}\ \Delta G^\circ_C &= -RT\ln K_p\\ &= -(1.987\ \text{cal}\ {}^\circ\text{K}^{-1}\ \text{mole}^{-1})\ (673^\circ\text{K})\ (2.303)\ \log(0.500)\\ &= 927\ \text{cal / mole.}\end{aligned}$$

$$\begin{aligned}\Delta G^\circ_P &= -RT\ln K_p\\ &= -(1.987\ \text{cal}\ {}^\circ\text{K}^{-1}\ \text{mole}^{-1})\ (673\ {}^\circ\text{K})\ (2.303)\ \log(1.64\times10^{-4})\\ &= 11{,}657\ \text{cal / mole.}\end{aligned}$$

ΔG°_C is the change in Gibbs free energy when 1 mole of N_2 in the ideal gas state at the concentration of 1 mole per liter and 3 moles of H_2 in the ideal gas state at 1 mole per liter react to form 2 moles of NH_3 in the ideal gas state at 1 mole per liter.

ΔG°_P is the change in Gibbs free energy when 1-mole of N_2 in the ideal gas state at 1 atm, and 3 moles of H_2 in the ideal gas state at 1 atm react to form 2 moles of NH_3 in the ideal gas state gas state at 1 atm.

Q. 12 : Determine the equilibrium constant for the following reaction at 25°C :

$$Mg(s) + Sn^{2+} \rightarrow Mg^{2+} + Sn(s)$$

Solution: The equilibrium constant, k, is related to the cell potential, $\Delta E°$, by means of the following two formulas :

$\Delta G° = -nF\Delta E°$, and $\Delta G° = -2.303\ RT \log k$,

where n is the number of electrons transferred, F the Faraday constant, R the gas constant, $\Delta G°$ the free energy, and T the absolute temperature. $\Delta E°$ will be determined from the half-cell potentials, E°.

First, the cell potential is determined from the halfcell reactions :

$Mg\ (s) \rightarrow Mg^{2+} + 2e^-$ $E° = +2.37$ v

$Sn^{2+} + 2e^- \rightarrow Sn\ (s)$ $E° = -0.14$ v.

Adding the two half-cell reactions and their respective values for E°,

$Mg\ (s) + Sn^{2+} + 2e^- \rightarrow Mg^{2+}\ 2e^- + Sn(s)$

$\Delta E° = (2.37v) + (-0.14\ v)$

or, $Mg\ (s) + Sn^{2+} \rightarrow Mg^{2+} + Sn(s)$ $\Delta E° = 2.23$ v.

The cell potential for the reaction is therefore 2.23v.

Equating the two expressions

$\Delta G° = -nF\Delta E°$

and $\Delta G° = -2.30\text{-}3\ RT \log k$

gives $-2.303\ RT \log k = -nF\Delta E°$,

or, $$\log k = \frac{nF\Delta E°}{2.303RT}.$$

Since two electrons are transferred from Mg(s) to Sn^{2+} n = 2. The absolute temperature is T = 25°C + 273 = 298°K. Hence

$$\log k = \frac{nF\Delta E°}{2.303\ RT}$$

$$= \frac{(2)(2.30609 \times 10^4\ \text{cal/mole} - \text{v})(2.23\text{v})}{(2.303)(1.987\ \text{cal/mole-deg})\ 298°\text{k}} = 75.6,$$

or, $k = 10^{75.6} = 4 \times 10^{75}$.

The equilibrium constant for this reaction is therefore 4×10^{75}.

8

Chemical Kinetics and Photochemistry

Q. 1 : Under certain conditions, the rate equation for the formation of water

$$2H_2 + O_2 \rightarrow 2H_2O$$

is given by

$$\text{rate} = k\,[H_2]^2\,[H_2O]$$

where k is the rate constant. What is the overall order of this rate equation ?

Solution: The overall order of a rate equation is equal to the sum of the exponents to which the concentrations are raised. In the equation

$$\text{rate} = k\,[H_2]^2\,[H_2O],$$

$[H_2]$ is raised to the second power and $[H_2O]$ is raised to the first power. Hence, the rate is second order in $[H_2]$, first order in $[H_2O]$, and 2 + 1 = 3, or third order overall.

Q. 2 : The rection

$$A + B \rightarrow C$$

was studied kinetically and the following data was obtained.

Experiment	A	B	Rate (mole/liter-min)
1	1.0 M	1.0 M	0.15
2	2.0 M	1.0 M	0.30
3	1.0 M	2.0 M	0.15

Determine the rate expression.

Solution: The rate of reaction is equal to some rate constant, k, multiplied by the concentrations of A and B raised to the appropriate powers. That is,

$$\text{rate} = k\,[A]^m\,[B]^n$$

where the exponents m and n are to be determined.

Comparing experiments 2 and 1, we see that holding [B] constant while doubling [A] doubles the rate of reaction (from 0.15 to 0.30). Hence, the rate is directly proportional to [A] and m = 1. Thus, if we hold [B] constant and triple [A], the rate triples.

Comparing experiments 3 and 1, we see that holding [A] constant and changing [B] (from 1.0M to 2.0M) has no effect on the rate. Hence, the rate is independent of [B] and n = 0 (so that [B] = 1 and [B} does not appear in the rate expression).

Substituting m = 1 and n = 0 into the rate expression gives rate = k $[A]^1\,[B]^0$, or, rate = k [A].

$0.030\ \text{M/min} = k\,(0.10\ \text{M})^2\,(0.10\ \text{M})$

$$\frac{0.030\ \text{M}/\text{min}}{(0.10\ \text{m})^2(0.10\ \text{M})} = k$$

$$30.0\ \text{M}^{-2}\ \text{min}^{-1} = k$$

(c) One can solve for the initial rate of the formation of Z by using the rate law and k when [X] and [Y] are given. Thus,

$$\text{Rate} = k[X]^2\ [Y]$$

$$\text{Rate} = (30.0\ \text{M}^{-2}\ \text{Min}^{-1})\ (0.15\ \text{M})^2\ (0.15\ \text{M})$$

$$= .101\ \text{M}/\text{min}.$$

(d) According to Boyle's Law the pressure is inversely proportional to the volume. Therefore, if the volume is doubled the pressure is halved. Because the reactants in this system are gases, when the pressure is halved, the concentration is halved. One can find the rate of the reaction described in (c) by using the rate law with the concentrations of the reactants halved

$$\text{Rate} = 30.0\ \text{M}^{-2}\ \text{min}^{-1}\ (½(0.15\ \text{m}))^2\ ½(0.15\ \text{M})$$

$$= 30.0\ \text{M}^{-2}\ \text{min}^{-1}\ (.075\ \text{M})^2\ (.075\ \text{M})$$

$$= 1.27 \times 10^{-2}\ \text{M}/\text{min}$$

The percent of the first rate that the second rate is equal to is found by dividing the second rate by the first and then multiplying the quotient by 100.

$$\frac{\text{Rate}_2}{\text{Rate}_1} \times 100 = \frac{1.27 \times 10^{-2}}{1.01 \times 10^{-1}} \times 100 = 12.5\% \text{ or } 1/8.$$

The second rate is, therefore, 1/8 of the first.

Q. 3 : Photochemical smogs are thought to be partially initiated by the photolytic decomposition of NO_2 to form NO and O. When the subsequent reaction

$$2NO + O_2 \rightarrow 2NO_2$$

is studied kinetically, it is found that when the initial concentration of O_2 is doubled and that of NO held constant, the initial reaction rate doubles; and when the initial concentration of NO is doubled and that of O_2 held constant, the initial reaction rate quadraples. Write the rate expression for this reaction.

Solution: The rate of reaction is equal to some rate constant k multiplied by the concentrations of NO and O_2 raised to the appropriate powers. That is,

$$\text{rate} = k\,[NO]^m\,[O_2]^n$$

where the exponents m and n are to be determined.

When [NO] is held constant and $[O_2]$ is doubled, the rate is doubled. Hence, the rate is directly proportional to $[O_2]$ and n = 1. Thus, if we were to hold [NO] constant and triple $[O_2]$, the rate is tripled.

When $[O_2]$ is held constant and [NO] is doubled, the reaction rate is quadrupled. Hence, the rate is second order in [NO] and m = 2. Thus, if we were to hold $[O_2]$ constant and triple [NO], the rate is multiplied by $3^2 = 9$.

Substituting m = 2 and n = 1 into the rate expression above gives

$$\text{rate} = k\,[NO]^2\,[O_2]^1, \qquad \text{or, rate} = k\,[NO]^2\,[O_2].$$

Q. 4 : What is the half-life of an unstable substance if 75% of any given amount of the substance decomposes in one hour ?

Solution: The half-life is defined as the time it takes for one half of the amount of a substance to decompose. When given the time elapsed and the percent of decomposition, one can find the half-life. One knows that ½ of the substance decomposes in the time equal to the half-life. This leaves ½ of the substance, ½ of this decomposes in the next span of time elapsed equal to the half-time. This leaves 1/4 of the substance. 75% of it has decomposed after two half-lifes have elapsed. Thus 2 half-lifes equal one hour or the half-life or the substance is ½ hour.

Q. 5 : For the first order decomposition of azomethane at 600°K it takes 30 minutes for the original concentration to decrease to half its value. After 60.0 minutes have elapsed, what percentage of the azomethane originally present remains ?

Solution: The decomposition is first order which means rate = k [A], where k is a constant and [A] is the concentration of azomethane. The half-life is independent of time, i.e., it is a constant and is defined as the time necessary for half the particular reactant present to decompose. Thus, if you knew the half-life of azomethane, you could calculate the percentage of azomethane that remains after a certain period of time. The half-life of azometane is 30 minutes. Thus, after 30 minutes, you have half of the original material left. After 60 miniutes half of this original material is left of 1/4 of the original material remains.

Q. 6 : When heated to 600°C, acetone (CH_3 $COCH_3$) decomposes to give CO and various hydrocarbons. The reaction is found to be first order in acetone concntration with a half-life of 81 sec. Given at 600°C a 1-liter container into which acetone is injected at 0.48 atm, approximate how long would it take for the acetone pressure to drop to 0.45 atm ?

Solution: When a reaction is said to be first order in a particular reactant it means that the rate of the reaction is proportional to the concentration of that reactant. Thus the rate law for this rection can be written :

$$\text{Rate} = k\,[CH_3COCH_3]$$

where k is the rate constant and [ ] indicate concentration. The rate constant is related to the half-life ($t_{½}$) by the following equation :

$$t_{½} = \frac{0.693}{k}$$

Solving for k : $t_{½}$ = 81 sec

$$81 \text{ sec} = \frac{0.693}{k}$$

$$k = 8.56 \times 10^{-3}\ \text{sec}^{-1}$$

To solve for the time it takes for a reaction to proceed to a certain degree one uses the equation :

$$d\ell n[A] = -k\,dt$$

where $d\ell n[A]$ is the natural logarithm of the original concentration subtracted fromt he natural log of the final concentration of acetone, k is the rate contant and dt is the time elapsed. Because acetone is a gas in this experiment the pressure is proportional to the concentration. Therefore one can let $[A]_{original} = 0.48$ and $[A]_{final} = 0.45$. Solving for dt :

$$\ell n\ 0.48 - \ell n\ 0.45 = (-8.56 \times 10^{-3}\ sec^{-1})\ (dt)$$

$$(-.734) - (-.799) = (-8.56 \times 10^{-3}\ sec^{-1})\ (dt)$$

$$\frac{(-.734) - (-.799) = dt}{-8.56 \times 10^{-3}\ sec^{-1}}$$

$$-7.5\ sec = dt$$

Therefore 7.5 sec have elapsed since the beginning of the reaction.

Q. 7 : What activation energy should a reaction have so that raising the temperature by 10°C at 0°C would triple the reaction rate ?

Solution: The activation energy is related to the temperature by the Arrhenius equation which is stated

$$k = Ae^{-E/RT}$$

where A is a constant characteristic of the reaction; e is the base of natural loragithms, E is the activation energy, R is the gas constant ($8.314\ J\ mol^{-1}\ deg^{-1}$) and T is the absolute temperature. Taking the natural log of each side:

$$\ell n\ k = \ell n\ A - E/RT$$

For a reaction that is 3 times as fact, the Arrhenius equation becomes

$$3\,k = Ae^{-E/R(T+10^{\circ})}$$

Taking the natural log :

$$\ell n\ 3 + \ell n\ k = \ell n\ A - E/R(T+10^{\circ})$$

Subtracting the equation for the final state from the equation for the initial state:

$$\ell n\ k = \ell n\ A - E/RT$$

$$-(\ell n\ 3 + \ell n\ k = \ell n\ A - E/R(T+10))$$

$$-\ell n\ 3 = -E/RT + E/R(T+10)$$

Solving for E:

$- \ell n\ 3 = - E/RT + E/R\ (T + 10)$ $\qquad R = 8.314$ J/mole °K

$T = 0 + 273 = 273$

$- \ell n\ 3 = - E/8.314$ J/mole-K) (273 K) + E/(8.314 J/mole–K) (283K)

$- 1.10 = - E/2269.72$ J/mole + E/2352.86 J/mole

(2269.72 J/mole) (2352.86 J/mole) × – 1.10 =

(–E/2269.72 J/mole + E/2352.86 J/mole) (2269.72 J/mole) (2352.86 J/mole)

-5.874×10^{6} J²/mole² = (–E) (2352.86 J/mole) + E (2269.72 J/mole)

-5.874×10^{6} J²/mole² = $- 8.314 \times 10^{1}$ J/mole × E

7.06×10^{4} J/mole = E

Q. 8 : Determine a and b in the rate equation

$$\text{rate} = kC_A^a C_B^b$$

given the following:

Table 1:

rate, M s⁻¹	0.05	0.10	0.20	0.40
$C_{A,0,}$M	1	1	2	2
$C_{B,0,}$M	1	2	1	2

Calculate k for this reaction.

Solution: log rate = log k + n log C_i (1)

where k = rate constant

n = constant, called order of reaction (a and b in this case)

C_i = concentration of substance i

Using equation (1), the order for both substances A and B can be determined. Using data from the first and third columns of table 1 and substituting the respective values into equation (1), gives

$$\log (0.05) = \log k_A + a \log 1$$

and

$$\log (0.20) = \log k_A + a \log 2,$$

keeping ($C_{B,0}$) fixed.

If the first equation is subtracted from the second, then

$\log(0.20) - \log(0.05) = a \log 2 - a \log 1$
$\log(0.20/0.05) = a(\log(2/1))$
$0.602 = a(0.301)$

$$\therefore \quad a = \frac{0.602}{0.301} = 2$$

Doing the same for substance B, using the first and second columns and keeping $C_{A,0}$ fixed, gives

$\log(0.05) = \log k_B + b \log 1$

and

$\log(0.10) = \log k_B + b \log 2$

If the first equation is subtracted from the second, then

$\log(0.10) - \log(0.05) = b \log 2 - b \log 1$

$$\therefore \quad \log = \frac{0.10}{0.05} = b \log(2/1)$$

$(0.301) = b(0.301)$

$$\therefore \quad b = \frac{0.301}{0.301} = 1$$

Therefore, $\quad \text{rate} = kC_A^2 C_B$

Solying for k, yields $\quad k = \dfrac{\text{rate}}{C_A^2 C_B}$

Using values of C_A, C_B and rate from table 1

$$k = \frac{0.05}{(1^2)(1)} = \frac{0.10}{(1^2)(2)} = \frac{0.20}{(2^2)(1)} = \frac{0.40}{(2^2)(2)}$$

form which $k = 0.05\ M^{-2}s^{-1}$ (no matter which column is chosen).

Q. 9 : At 393.7°C, the second-rate constant for the decomposition of gaseous hydrogen iodide in a second order reaction is $2.6 \times 10^{-4} M^{-1}s^{-1}$. Given that the Arrhenius activation energy for the reaction is 45.6 kcal mol^{-1}, calculate the second order rate constant for this reaction from collision theory. The molecular weight of HI = 127.9g mol^{-1}, the collision diameter = 3.5 Å and the orientation factor, P = 1.

Solution: The collision theory postulates that is order for molecules to interact, they must approach each other so closely that they can be said to be in collision. The speed of the reaction is equal to the number of collisions per second times the fraction of the collisions that are effective in producing chemical change.

Thus, according to this gas-collision theory, the second order rate constant should be given by

$$k = 2 \times 10^3 N_A P \left(\frac{\pi RT}{M}\right)^{1/2} \sigma^2 e^{-(E_O/RT)} \quad (1)$$

where N_A = the Avogadro's number, P is the steric or orientation fractor, σ is the collision diameter and E_o is the threshold energy for the reaction, given by the expression

$$E_O = E_a - \frac{RT}{2} \quad (2)$$

E_a is the activation energy

Substitute the respective values into equation (2) to obtain

$$E_O = 45.6 \text{ kcal mole}^{-1}$$

$$- \frac{(1.987 \times 10^{-3} \text{kcal mole}^{-1} {}^\circ K^{-1})(666.7^\circ K)}{2}$$

$$= (45.6 - 0.662) \text{ kcal mole}^{-1}$$

$$= 44.94 \text{ kcal mole}^{-1}$$

Form equation (1),

$$k = 2 \times (10^3 \text{ liters m}^{-3})(6.02 \times 10^{23} \text{ mole}^{-1})(1)$$

$$X\left[\frac{\pi\left(8.31 \frac{J\,K^{-1}}{\text{mole}}\right)(666.7^\circ K)}{127.9 \times 10^{-3} \text{kg mole}^{-1}}\right]^{1/2}$$

$$x\left[3.5Å\left(\frac{10^{-10}\text{ m}}{1Å}\right)\right]^2 e^{-(44.94 \text{ kcal mole}^{-1})/(1.987 \text{ call } {}^\circ K^{-1}\text{mole}^{-1})}$$

$$x(666.7^\circ K)$$

$$= 1.0 \times 10^{-4} M^{-1}s^{-1}$$

Q. 10 : Measurement of the second order decomposition of acetaldehyde over the temperature range 700-850°K was made and the second order rate constant was given as :

T(°K)	700	730	760	790	810	910	1000
k_2	0.011	0.035	0.105	0.343	0.789	2.17	145

Determine by a plot the activation energy and the pre-exponential factor for the decomposition.

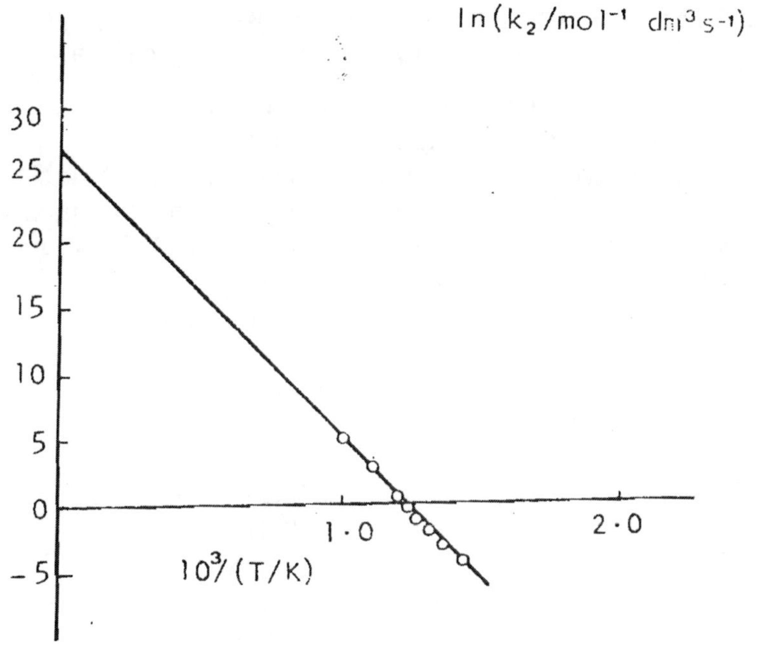

Solution: The variation of temperature with reaction rate constants is usually expressed by the Arrhenius equation,

$$k_2 = Ae^{-E_a/RT} \qquad (1)$$

where A is the pre-exponential factor and E_a is the activation energy.

Equation (1) can be written in a logarithmic form as follows:

$$\ln k_2 = -\frac{E_a}{R}\frac{1}{T} + \ln A \qquad (2)$$

Form equation (2), it can be seen that a straight line will be obtained if the logarithm of the rate constant is plotted against the reciprocal of the absolute temperature.

From the plot, the slope is determined to be -2.207×10^4 °k

Thus $E_a = -[(8.314 \text{ J K}^{-1} \text{ mole}^{-1})(-2.207 \times 10^4 \text{ °K})]$

$= 1.84 \times 10^5$ J mole^{-1}

$= 184$ kJ mole^{-1}

The intercept on the plot has a value of 26.95 and thus

$A = 5.06 \times 10^{11}$.

Q. 11 : Chloroform reacts according to the following equation in the presence of sodium methoxide in methanol solution.

$$CHCl_3 + OMe^- \rightleftarrows CCl_3^- + MeOH \text{ (fast)}$$

$$CCl_3^- \rightarrow CCl_2 + Cl^- \text{ (slow)}$$

and $$CCl_2 \rightarrow \text{products (fast)}$$

An experiment was done to measure the order of the slow step with respect to CCl_3^- and to determine the amount of Cl^- produced in time, t at 59.7°C by titration with 0.0100 N $AgNO_3$, V, ml being required. The following results were obtained.

t (min)	V, ml
0	1.71
4	3.03
9	4.49
15	5.97
22	7.39
30	8.87
41	10.48
50	11.7
∞	15.98

Show that the reaction is first order and determine the rate constant.

Solution: One way to determine the order of the reaction is to plot log $(V_\infty - V)$ vs t. A straight line would be obtained if the reaction is first order. From the slope of the plot, the value of k can be calculated.

Another method is to use the equation

$$k = \frac{2.303}{t} \log \frac{V_\infty - V_0}{V_\infty - V}$$

For $t = 4$ min

$$k = \frac{2.303}{4 \text{ min}} \log \frac{15.98 - 1.71}{15.98 - 3.03}$$

$$= 0.0243 \text{ min}^{-1}$$

For $t = 4$ min to 5 = 50 min

k = 0.0243, 0.0241, 0.0236, 0.0231, 0.0232, 0.0233 and 0.0241

Since k shows constancy then it can be said that the reaction is first order. And the value for k, taken as an average of all the values, is approximately 0.0237.

Q. 12 : The following mechanism is proposed for the photodimerization of substance A:

$$A + h\nu \xrightarrow{K_1} A^*$$

$$A^* + A \xrightarrow{K_2} A_2$$

$$A^* \xrightarrow{k_3} A + h\nu'$$

Here, h = Planck's constant and
v = Frequency of incident light

Derive the expression for the quantum yield of A_2.

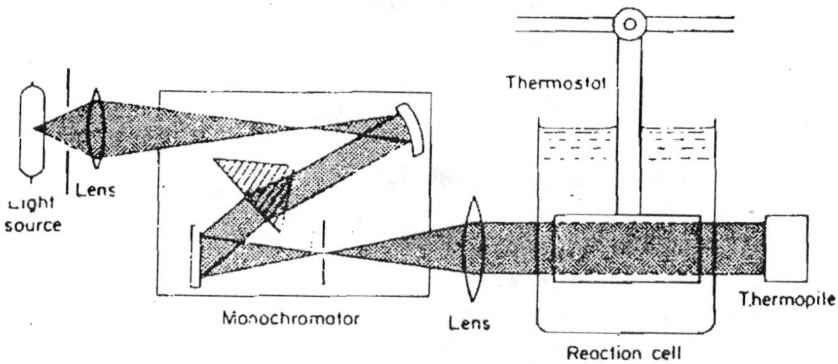

Arrangement for photochemical investigations.

Solution: The quantum yield of A_2 is the number of molecules of reactants consumed or product generated per quantum absorbed. That is

$$\frac{d(C_{A_2})/dt}{I}$$

where I = intensity of the light absorbed. So, beginning with the expression for the rate of change of the concentration of A* gives

$$\frac{dC_{A^*}}{dt} = k_1 I - k_2 C_{A^*} C_A - k_3 C_{A^*} \qquad (1)$$

where C = concentration

k_1, k_2 and k_3 = rate constants

Assuming steady-state conditions, equation (1) is set equal to zero because the steady-state approximation assumes that the concentrrations of certain intermediate substances reach constant values. That is

$$\frac{dC_{A\bullet}}{dt} = k_1 I - k_2 C_{A\bullet} C_A - k_3 C_{A\bullet} = 0 \tag{2}$$

Rearranging equation (2) to solve for $C_{A\bullet}$ yields

$$C_{A\bullet} = \frac{k_1 I}{k_2 C_A + k_3} \tag{3}$$

The expression for the formation of A_2 is given by

$$\frac{d}{dt}\left(C_{A_2}\right) = k_2 C_{A\bullet} C_A \tag{4}$$

The $C_{A\bullet}$ term in equation (4) is given by equation (3). Substituting this expression into equation (4) gives

$$\frac{d\left(C_{A_2}\right)}{dt} = k_2 \frac{k_1 I}{k_2 C_A + k_3} C_A \tag{5}$$

Recall that the quantum yield is

$$\left(\frac{dC_{A_2}}{dt}\right) / I.$$

Hence, equation (5) changes to

$$\left(\frac{dC_{A_2}}{dt}\right) / I = \frac{k_2 k_1 C_A}{k_2 C_A + k_3}.$$

9

Solid State

Q. 1 : Sodium crystallizes in the body-centered cubic structure with $a = 4.24$ Å. Calculate the theoretical density of Na.

Solution: The content of a body-centered unit cell is $8(1/8) + 1 = 2$. In a body-centered cell, there are eight corners and their contribution is $8(1/8) = 1$. In other words, every corner of the unit cell contributes 1/8 of an atom. Also there is an atom on the body, which is not shared, making it 2 altogether.

$$d = \text{density} = \frac{ZM}{AV} \qquad (1)$$

where Z = content = 2

M = Molecular weight

A = Avogadro's number

$V = abc\,(1 - \cos^2\alpha - \cos^2\beta - \cos^2\gamma + 2\cos\alpha\cos\beta\cos\gamma)^{1/2}$

But for a cell having 90° (as in this case), $V = abc$. a, b and c are lengths of the edges. V = Volume. Therefore

$$V = (4.24\text{Å})^3 = (4.24 \times 10^{-10}\text{m})^3$$

using (1),

$$d = \frac{(2)\left(23\ \text{gm mol}^{-1}\right)\left(10^{-3}\text{kg gm}^{-1}\right)}{\left(6.022 \times 10^{23}\text{mol}^{-1}\right)\left(4.24 \times 10^{-10}\text{m}\right)^3}$$

$$= 1.00 \times 10^3\ \text{kg m}^{-3}.$$

Q. 2 : Determine the coordinate number for an atom in the body-centered cubic unit cell.

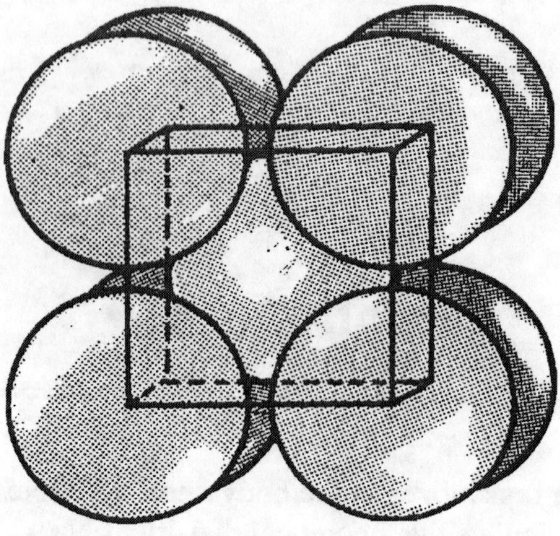

Fig.1

Solution: Fig. 1 is an example of a body centered cubic cell. Now consider the atom in the center, it is surrounded by eight atoms. These are called the nearest neighbor atoms. In other words, the atom in the center of a body centered cubic unit cell is surrounded by eight nearest neighbor atoms.

By definition, the coordinate number of an atom in a crystal is the number of nearest-neighbor atoms. Therefore, a body centered cubic unit cell has a coordinate number of 8.

Q. 3 : Excluding hexagonal unit cells, when counting the number of points inside a cell, a point on an edge is 1/4 inside the cell, and a lattice point at a corner is 1/8 inside the cell. Justify these fractions. Calculate, also, the net number of lattice points in the following unit cells: simple cubic, body-centered cubic, face-centered cubic, and tetragonal.

Solution: To solve this problem properly you need to know the definition of a unit cell and lattice point. You need, also, to know the actual structures of the unit cells given. You proceed as follows: A unit cell is that small fraction of a space lattice, which sets the pattern for the whole lattice. In other words, it is the smallest portion of the space lattice (which is just the pattern of points which describes the arrangement of atoms or molecules in the crystal), which, when moved repeatedly a distance equal to its own dimensions alogn the various directions, generates the whole space lattice equivalent to the original lattice. The lattice is, of course, just points that denote atoms or molecules; thus, the term lattice point. The accompanying illustrations give you the structures in question. From this, you can obtain justification for the fractions and find the net number of lattice points. For example, take the simple cubic structure. In a space lattice, each atom is shared with three other unit structures as seen in figure d.

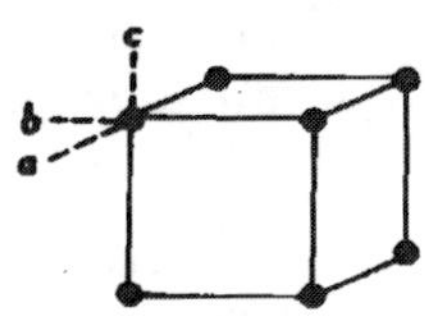

simple cubic
(a)

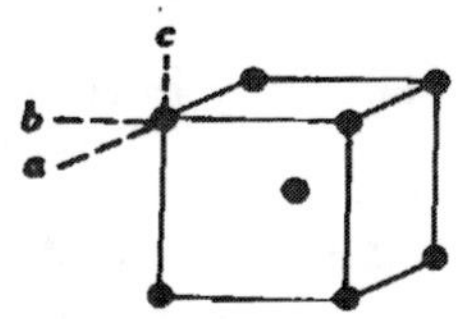

body-centered cubic
(b)

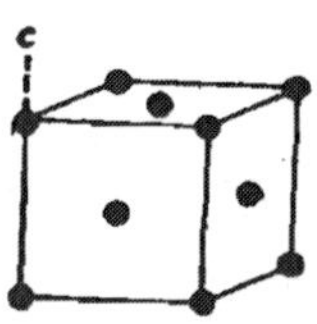

face-centered cubic
(c)

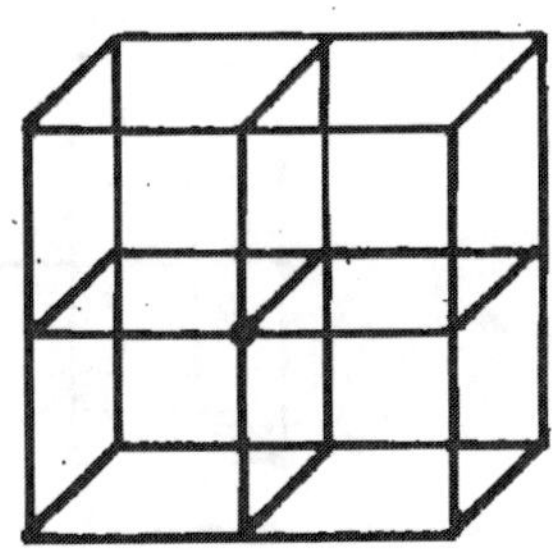

(d)

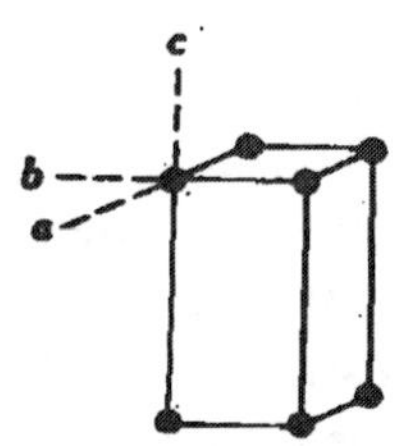

tetragonal
(e)

Note: This point is shared by 8 unit cells, if you place another 4 cubes on the side. Thus, each point is only 1/8 in the cell. This same sort of procedure and reasoning can be used to justify all fractions in all unit cells given. Calculations now become easy. Simple cubic: Have 8 lattice points. Each is only 1/8 inside cell. Thus, you have net of (1/8) (8) = 1.

Face-centered cubic: 8 lattice points at corners, each only 1/8 inside cell. Have a point on each face. Each point is only ½ inside cell. Have six points. Thus, net = (8) (1/8) + (6) (½) = 4.

Body-centered: Again have (8) (1/8) from lattice points at corners. 1 point in center, which is entirely in cell. Thus net becomes (8) (1/8) + (1) (1) = 2.

Tetragonal: Only have a net of (8) (1/8) from lattice points at corners.

Q. 4 : Metallic gold crystallizes in the face-centered cubic lattice. The length of the cubic unit cell, a, is 4.070Å. (a) What is the closest distance between gold atoms ? (b) What is the density of gold ?

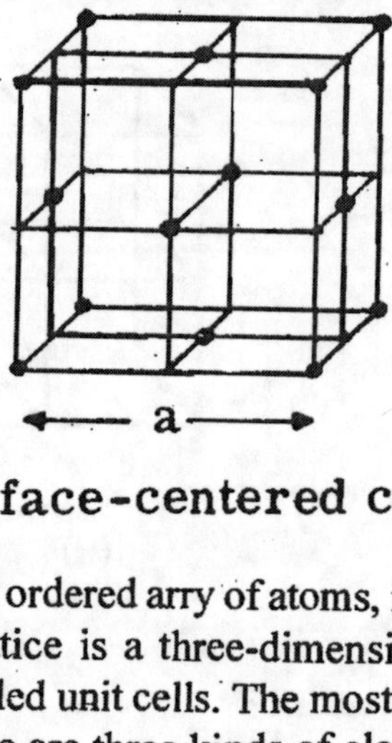

face-centered cubic

Solution: An ordered arry of atoms, ions, or molecules is called a lattice. Every lattice is a three-dimensional stacking of identical building blocks called unit cells. The most symmetrical crystals have cubic lattices. There are three kinds of elementary cubic lattices: the simple cubic, the face-centered cubic, and the body-centered cubic. The lenght of the cube edge is designated by the symbol a. In a face-centered lattice, an atom would be located at each corner of the unit cell and at the center of each of the six faces.

(a) In the gold crystal above, the closest distance from one corner atom to another corner atom is a, 4.070Å. The distance from any corner atom to an atom at the center of a face is one-half the diagonal of that face. Using the fact that the diagonal of any square is $\sqrt{2} \times$ length of one side, the diagonal is then $a\sqrt{2}$. Half the diagonal is ½ $a\sqrt{2}$ or

$$½(4.070Å)\ \sqrt{2} = 2.878\ Å$$

This distance is the closest distance between atoms, since this distance is shorter than the distance between adjacent corners.

(b) The density is computed by first counting the number of gold atoms that can occupy one unit cell. This number is 1/8 times the number of occupied corners in the unit cell plus ½ times the number of occupied face-centers. Since a cube has 8 corners and 6 faces, the number is

$$1/8.\ 8 + ½\ .\ 6 = 4$$

The volume of the cubic unit cell is a^3. The density if the mass of 4 gold atoms (4 atomic Wt/Avogadro's #) divided by the volume of the unit cell, a^3. Thus the formula for the density of a gold crystal lattice is

$$\text{Density} = \frac{(4\ \text{atoms}).(\text{Atomic Wt. of gold})}{a^3\ (\text{Avogadro's number})}$$

Substituting into the equation the Atomic Weight of gold, 197.0 g/mole, Avogadro's number, 6.023×10^{23} atoms/mole, and converting 4.070Å to 4.070×10^{-8} cm, then

$$\text{Desnsity} = \frac{(4\ \text{atoms})\ (197.0\ \text{g/mole})}{(4.070 \times 10^{-8}\text{cm})^3\ (6.023 \times 10^{23}\ \text{atoms/mole})}$$

$$= 19.4\ \text{g/cm}^3.$$

Q. 5 : Find the ionic radius of I^-, given that the unit-cell edge length of LiI is measured to be .6240 nm. Assume that the large, negative ions are in actual contact with the diagonal.

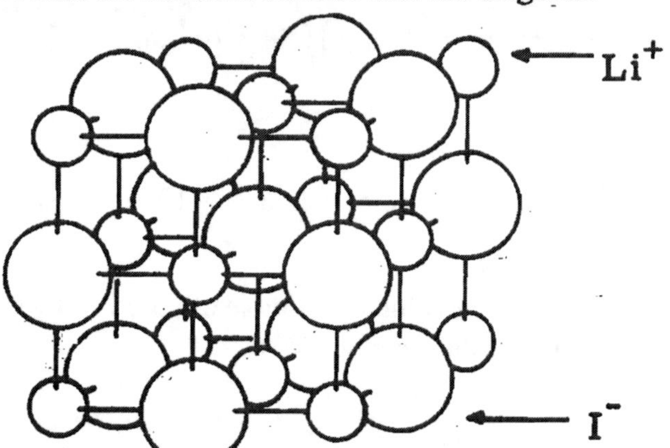

Fig. The Structure of Lithium Iodide

Solution: To solve this problem, determine the length of the face diagonal in LiI space lattice, which is the pattern of the arrangement or atoms of molecules in a crystal. Because LiI has the same form of the cubic unit cell as NaCl, its space lattice will possess 3 I^- ions on the face diagonal. Thus, if one knows the length of the face diagonal, one can divide by 4 to deduce the radius of I^-. The face diagonal can be found (from the Pythagorean Theorem) by multiplying ($\sqrt{2}$) times (cubic-edge length). Thus, the face diagonal = ($\sqrt{2}$) (.6240) = .8823 nm. However, face diagonal = 4 I^- radii.

$$\text{Thus, } I^- \text{ radius} = \frac{.8823}{4} = .2206 \text{ nm.}$$

Q. 6 : The molecular diameter of an N_2 molecule, as deduced from the Van der Waals b parameter, is 3.15×10^{-8} cm. The density of liquid nitrogen is 0.8081 g/cm³. On a hard-sphere model, what fraction of the liquid volume appears to be empty space ?

Solution: To find the fraction of the liquid volume that appears to be empty space, one must subtract the volume of spheres from the total volume of the liquid. One can find the volume of the spheres that constitute one mole of N_2 liquid by multiplying the volume of one sphere by 6.02×10^{23}, the number of spheres in one mole. The volume of a sphere is equal to 4/3π r3 where r is the radius of the sphere. The diameter of the N_2 sphere is given as 3.15×10^{-8} cm, thus the radius is half of length or 1.58×10^{-8} cm. Solving for the volume of 1 sphere :

$$\text{Volume of 1 sphere} = 4/3\,\pi\left(1.58 \times 10^{-8}\,\text{cm}\right)^3$$

$$= 1.652 \times 10^{-23}\,\text{cm}^3/\text{sphere}$$

Volume of 1 mole of spheres

$$= \left(6.02 \times 10^{23}\,\text{spheres/mole}\right) \times \left(1.652 \times 10^{-23}\,\text{cm}^3/\text{sphere}\right)$$

$$= 9.945\,\text{cm}^3/\text{mole}$$

One can find the total volume of 1 mole of liquid N_2 by dividing the molecular weight by the density. (MW = 28)

$$\text{total volume} = \frac{28\,\text{g/mole}}{0.8081\,\text{g/cm}^3} = 34.65\,\text{cm}^3/\text{mole}.$$

The volume of the empty space is equal to the volume of the

spheres subtracted from the total volume.

volume of empty space per mole = 34.65 cm³ – 9.945 cm³

= 24.70 cm³

The percent of the volume taken up the empty space is equal to the volume of the space divided by the total volume and multiplied by 100.

$$\% \text{ of empty space} = \frac{24.70 \text{ cm}^3}{34.65 \text{ cm}^3} \times 100$$

$$= 71.30\%.$$

Q. 7 : Find the spacing between the planes with indices 101 in NaCl if a = 5.6402 Å.

Solution: By definition, $\frac{1}{d^2_{hkl}} = \frac{h^2 + k^2 + l^2}{a^2}$ (1)

where d = spacing

hkl = Miller indices (101 respectively in this case). Substituting the respective values into (1), yields

$$\frac{1}{d^2_{101}} = \frac{1^2 + 0^2 + 1^2}{(5.6402\text{Å})^2} = \frac{2}{31.81\text{Å}^2}$$

$$d^2_{101} = \frac{31.81\text{Å}^2}{2}$$

$$d_{101} = \sqrt{\frac{31.81\text{Å}^2}{2}} = \frac{5.6402\text{Å}}{1.4142} = 3.9883\text{Å}.$$

Note that the general formula is

$$\frac{1}{d^2_{hkl}} = \frac{h^2}{a^2} + \frac{k^2}{b^2} + \frac{l^2}{c^2};$$

but in the problem, a = b = c.

Q. 8 : Determine the content for the body-centered cubic unit cell shown in the Fig. below.

Solution: There are eight corners in a body-centered cubic cell. The contribution of one corner is 1/8 of an atom. Therefore, the eight corners contribute

$$\frac{1}{8} + \frac{1}{8} + \frac{1}{8} + \frac{1}{8} + \frac{1}{8} + \frac{1}{8} + \frac{1}{8} + \frac{1}{8} = 8\left(\frac{1}{8}\right) = 1.$$

We can also see that there is one atom in the center unshared by any corner. Consequently, there are

$$\left[8\left(\frac{1}{8}\right)+1\right] \text{atoms} = 2 \text{ atoms}$$

in a body-centered cubic unit cell.

Q. 9 : Polonium is the only element known to crystallize in a primitive cubic unit cell under room conditions. If a = 3.36 Å, find the theoretical density.

Solution: Density, $d = \frac{ZM}{AV}$.

Z = content = 1

M = Molecular weight

Ǎ = Avogadro's number

V = Volume = a^3 in this case.

But V = $(3.36 \text{ Å})^3 = (3.36 \times 10\text{–}10\text{m})^3$

$$d = \frac{(1)\,(209\ \text{gm/mole})\,(10^{-3}\ \text{kg/gm})}{(6.022 \times 10^{23}\ \text{mole}^{-1})\,(3.36 \times 10^{-10}\ \text{m})^3}$$

$$= 9.15 \times 10^3\ \text{kg/m}^3.$$

Note that a primitive unit cell has points only at the corners of a parallelepiped. So the total contribution from all corners = 1. This is why Z = content = 1, unlike others like edge-entered, body-centered or face centered cells that have points on the edge, body and face respectively (as suggested by their names.)

Q. 10 : Find the distance between two Po atoms that lie along a body diagonal if

a = 3.36 Å.

Solution: Po atoms form a cubic unit cell. The distance, L, befween two atoms in a cubic unit cell is

$$L = \sqrt{a^2(x_2 - x_1)^2 + a^2(y_2 - y_1)^2 + a^2(z_2 - z_1)^2} \quad (1)$$

where $(x_1y_1z_1)$ and $(x_2y_2z_2)$ are their respective coordinates. In this case $(x_1y_1z_1)$ = (000) and $(x_2y_2z_2)$ = (111). Substituting these values into equation (1) gives

$$L = \sqrt{(3.36)^2\left[(1-0)^2 + (1-0)^2 + (1-0)^2\right]} = 3.36\text{Å}\sqrt{3}$$

$$= (3.36\text{Å})\ (1.732) = 5.8197\text{Å} = 5.820\text{Å}.$$

Q. 11 : Prepare a crystallographic projection for the body-centered cubic unit cell.

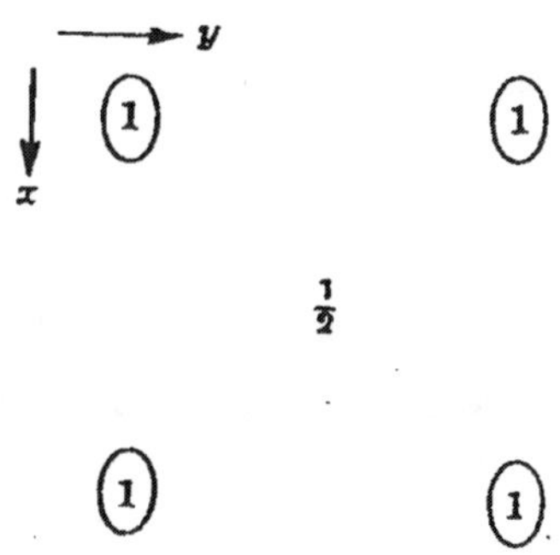

Solution: The eight corner atoms will be represented by four ① symbols at the corners of a square and the body-centered atom will be represented by a ½ in the center of the square, see Fig. below. The above is constructed by viewing a unit cell along one of the crystallographic axes. It shows the shape of a unit cell in two dimensions, with the three-dimensional information.

Q. 12 : If λ = 1.5418 Å for filtered Cu radiation, at what angle would the maximum reflection by the (200) plane of AgCl occur, assuming that

$$a = 5.5491\ \text{Å}?$$

Solution: A beam of X-rays will be reflected if

$$\sin\theta = \frac{n\lambda}{2d} = \frac{\lambda}{2d_{hkl}} \tag{1}$$

where θ = angle of reflection from the hkl plane.
n = order of reflection (an integer).
λ = wavelength of the radiation.
d_{hkl} = distance.

Looking at equation (1), $n = 1$ and $\lambda = 1.5418$ Å; but d_{hkl} is not given. This can be calculated using

$$\frac{1}{d_{hkl}^2} = \frac{h^2 + k^2 + l^2}{a^2}$$

where hkl = (200).
Therefore

$$\frac{1}{d_{200}^2} = \frac{2^2 + 0^2 + 1^2}{(5.5491\ \text{Å})^2} = 0.1299\text{Å}^{-2}$$

and

$$\frac{1}{d_{200}} = 0.3604\ \text{Å}^{-1}.$$

Thus

$$d_{200} = \frac{1}{0.3604\ \text{Å}^{-1}} = 2.7746\ \text{Å}.$$

Now that d_{hkl} is known, substitute its value into equation (1) and solve for θ. Hence

$$\sin\theta = \frac{1.5418\ \text{Å}}{(2)\ (2.7746\ \text{Å})} = \frac{1.5418}{5.5492} = 0.279$$

and

$$\theta = \sin^{-1}\ 0.279.$$

$$\theta = 16.1°.$$

10

Quantum Chemistry

Q. 1 : Determine the radii of the first and second quantum orbits in the hydrogen atom.

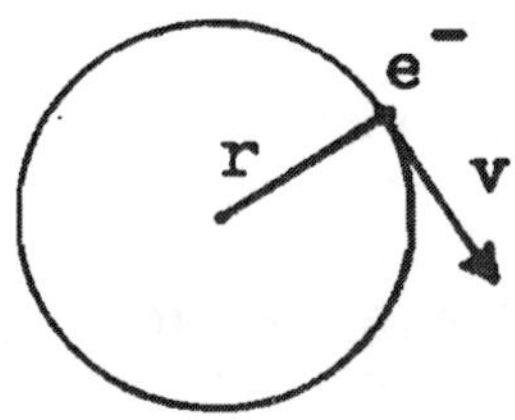

Solution: We will derive the Bohr theoretical expresion for the radius of the quantum orbit of an electron.

Consider an electron of mass m and charge a moving at a constant distance r from the nucleus with a tangential velocity v, as shown in the accompanying figure. In order for this orbit to be stable, the Coulombic force attracting the electron to the nucleus (e^2/r^2) must be balanced by the centrifugal force on the electron (mv^2/r), which, if not compensated for, would compel the electron to leave its orbit. Hence, the magnitude of these two forces must be equal.

$$\frac{e^2}{r^2} = \frac{mv^2}{r}$$

Solving for r, we obtain $r = \dfrac{e^2}{mv^2}$.

Bohr introduced quantization of the electron's orbit by assuming that the electron's angular momentum (mvr) is given by :

$$mvr = n\left(\frac{h}{2\pi}\right),$$

where h is Planck's constant and n is an index, called the principal quantum number, which takes on the values n = 1 for the first orbit, n = 2 for the second orbit, and so on. Solving for v,

$$v = \frac{nh}{2\pi mr}$$

We can now obtain an expression for r which includes quantization and is in terms of known quantities. Using the above expressions for v and r,

$$r = \frac{e^2}{mv^2} = \frac{e^2}{m}\left[\frac{2\pi mr}{nh}\right]^2 = \frac{4\pi^2 e^2 mr^2}{n^2h^2},$$

$$r = \frac{n^2h^2}{4\pi e^2 m}.$$

The radius of the first orbit (n = 1), is called the Bohr radius. Solving for r : e = 4.80×10^{-10} esu and m = 9.11×10^{-28} g.

$$r_1 = \frac{1^2 \times h}{4\pi^2 e^2 m} = \frac{1 \times (6.626 \times 10^{-27} \text{erg} - \text{sec})^2}{4\pi^2 \times (4.80 \times 10^{-10} \text{esu})^2 \times 9.11 \times 10^{-28} \text{g}}$$

$$= 5.29 \times 10^{-9} \text{ cm} = 0.529 \text{ Å}.$$

Let the radius of the nth orbit be denoted r_n. Then

$$r_n = \frac{n^2h^2}{4\pi^2e^2m} = n^2\left[\frac{h^2}{4\pi^2e^2m}\right] = n^2r_1 = n^2 \times 0.529 \text{ Å}.$$

Hence, the radius of the second quantum orbit is

$$r_2 = n^2 \times 0.529 \text{ Å} = (2)^2 \times 0.529 \text{ Å}$$

$$= 2.12 \text{ Å}.$$

Q. 2 : It has been suggested that the time required for light to travel 1 cm be used as a unit of time called the "jiffy". How many seconds correspond to one jiffy? The speed of light is about 3.0×10^8 m/sec.

Solution: The relationship between distance, speed, and time is

$$\text{distance} = \text{speed} \times \text{time, or, time} = \frac{\text{distance}}{\text{speed}}.$$

The time it takes light to travel 1 cm is:

$$\text{time} = \frac{\text{distance}}{\text{speed}} = \frac{1 \text{ cm}}{3 \times 10^8 \text{ m/sec}}$$

$$= \frac{1 \text{ cm}}{3 \times 10^8 \text{ m/sec} \times 10^2 \text{ cm/m}} = \frac{1 \text{ cm}}{3 \times 10^{10} \text{ cm/sec}}$$

$$= 3.3 \times 10^{-11} \text{ sec.}$$

Hence, 1 jiffy = 3.3×10^{-11} sec.

Q. 3 : Consider a molecule of hydrogen at 478.15 K and moving with a velocity of 2×10^5 cm/sec. What is the de Broglie wavelength of this hydrogen atom ?

Solution: The de Broglie wavelength (λ) is related to the momentum of a particle by the de Broglie equation :

$$\lambda = h/mv,$$

where h is Planck's constant, m is the mass of the particle, and v is its velocity (momentum = mv). Calculate the de Broglie wavelength of the hydrogen atom by directly substituting into this equation.

The mass of a hydrogen atom is m = 3.34×10^{-24} g. Planck's constant is h = 6.626×10^{-27} erg-sec = 6.626×10^{-27} g-cm^2/sec. Hence, the de Broglie wavelength is :

$$\lambda = \frac{h}{mv} = \frac{6.626 \times 10^{-27} \text{g} - \text{cm}^2/\text{sec}}{3.34 \times 10^{-24} \text{ g} \times 2 \times 10^5 \text{ cm/sec}} = 9.9 \times 10^{-9} \text{cm}$$

$$= 9.9 \times 10^{-9} \text{cm} \times 10^8 \text{ Å/cm} = 0.99 \text{ Å}.$$

Q. 4 : The mass of an electron is 9.11×10^{-31} and Planck's constant $= 6.63 \times 10^{-34}$ J sec. How much energy does an electron possess in its five lowest states, if it is placed in a cube of dimension 50 nm ?

Solution: Wave function theory states that the permitted values of the energy of a particle in a box is given as

$$E = \frac{N^2 h^2}{8 \text{ m} L^2},$$

where E = energy, N = principal quantum number, h = Planck's constant, m = mass of particle, and L = length of box. Substitute in the known values and solve for E. The principal quantum number N corresponds to the energy levels of the electrons. The five lowest energy states are N = 1, N = 2, N = 3, N = 4, and N = 5, where N = 1 is the lowest and N = 5 the highest energy level.

Substituting:

$$E = \frac{N^2 h^2}{8m L^2} = \frac{N^2 \left(6.63 \times 10^{-34}\,\text{J sec}\right)^2}{8\left(9.11 \times 10^{-31}\,\text{kg}\right)\left(5 \times 10^{-8}\,\text{M}\right)^2}$$

$$= N^2 \left(2.41 \times 10^{-23}\right)$$

(Note: 50 nm = 5.0×10^{-23})

1st energy level (lowest) : $N^2 = 1$, $E = 2.41 \times 10^{-23}$ J
2nd energy level : $N^2 = 4$, $E = 9.64 \times 10^{-23}$ J
3rd energy level : $N^2 = 9$, $E = 2.16 \times 10^{-22}$ J
4th energy level : $N^2 = 16$, $E = 3.85 \times 10^{-22}$ J
5th energy level : $N^2 = 25$, $E = 6.025 \times 10^{-22}$ J.

Q. 5 : Find the wavelength of light required for the ionization of sodium atoms to occur. The ionization potential of sodium is 8.17×10^{-19}J.

Solution: The energy necessary to ionize a sodium atom is equal to the ionization potential. Here, one is looking for the wavelength of light, which, when flashed upon sodium, will possess an energy equal to the ionization potential. Wavelength is related to energy by the equation $E = hc/\lambda$, where E is the energy, h Planck's constant (6.626×10^{-34} J sec), c the speed of light (3.0×10^8 m/sec) and λ the wavelength. One can rewrite this equation as $\lambda = hc/E$ and then substitute the given to solve for λ.

$$\lambda = \frac{\left(6.626 \times 10^{-34}\,\text{J sec}\right)\left(3.0 \times 10^8\,\text{m/s}\right)}{\left(8.17 \times 10^{-19}\,\text{J}\right)}$$

$$= 2.43 \times 10^{-7}\,\text{m} = 2.43 \times 10^{-7}\,\text{m} \times \frac{1\,\text{Å}}{10^{-10}\,\text{m}} = 2430\,\text{Å}.$$

Q. 6 : Calculate the average separation of the electron and the nucleus in the ground state of the hydrogen atom.

Solution: The ground state wavefunction of a hydrogen atom is the wavefunction corresponding to the lowest energy. It is spherically symmetrical, and decays exponentially with the distance from the nucleus. This is represented by :

$$\psi = \left[\frac{1}{\pi a_0^3}\right]^{1/2} e^{-r/a_0} \tag{1}$$

where ψ = wavefunction

a_0 = a collection of fundamental constants. It has the dimensions of length and a magnitude of 53 pm. It is called the Bohr radius.

r = distance between the electron and the nucleus.

The problem requires the evaluation of the expectation of r.

ψ^* (p) ψ (P) dxdydz is the probability of a particle appearing at the point, p if the three-dimensional volume is dxdydz. The requirement that the particle must be some-where in the universe is the sum over the probabilities that it appears in every element dxdydz into which the universe can be divided. Therefore, the mean position is given by

$$<p> = \int \psi^*(p)\, p\psi\,(p)d\tau \tag{2}$$

where $d\tau$ is the infinitesimal region of finding the particle.

It is more convenient to work in polar coordinates and use $d\tau = r^2\, dr \sin\theta d\phi$, allowing r to range from 0 to ∞, θ from 0 to π, and ϕ from 0 to 2π. Therefore, equation (2) for the expectation value of r bcomes

$$<r> = \int_0^{\infty} dr \left[r^2\, r \left[\frac{1}{\pi a_0^3}\right] e^{-2r/a_0} \right] \int_0^{\theta} d\theta \sin\theta \int_0^{2\pi} d\phi \tag{3}$$

Since $\int_0^{\pi} \sin\theta\, d\theta \int_0^{2\pi} d\phi = 2(2\pi) = 4\pi$,

equation (3) becomes

$$4\pi \int_0^{\infty} r^3 \left[\frac{1}{\pi a_0^3}\right] e^{-2r/a_0} dr.$$

But $\frac{1}{\pi a_0^3}$ is a constant. Therefore,

$$<r> = \frac{4\pi}{\pi a_0^3} \int_0^{\infty} r^3\, e^{-2r/a_0} dr$$

$$= \frac{4}{a_0^3} \left[6 \left[\frac{a_0}{2}\right]^4 \right]$$

$$= \frac{3}{2} a_0.$$

Recall that $a_0 = 53$ pm.

Hence $<r> = \frac{3}{2}(53) = 79.5$ pm.

Q. 7 : An electron was accelerated through a potential difference of 1.00 ± 0.01 kV. What is the uncertainty of the position of the electron along its line of flight ?

Solution: If the momentum of a particle is precisely specified, the position is totally uncertain. This remarkable destruction of one of the fundamentals of classical mechanics is a special case of the Heisenberg Uncertainty Principle. It states that if the momentum is known to lie within a range δp, then the position must be uncertain to an extent δz, where

$$\delta p \delta z \geq \frac{\hbar}{2} \quad (1)$$

Here $\hbar = \frac{h}{2\pi}$ and h = Planck's constant. Note that z is the line of flight. The uncertainty of momentum can be calculated from

$$\delta p = m_e \, \delta v \quad (2)$$

where δv = the uncertainty in velocity. The δv can be calculated from $\frac{1}{2} m_e v^2 = eV$.

For a potential difference of 0.99 kV,

$$v = \sqrt{\frac{2eV}{m_e}} = \sqrt{\frac{2(1.602 \times 10^{-19} C)(990 V)}{9.110 \times 10^{-31} kg}}$$

$$= \sqrt{\frac{2(1.602 \times 10^{-19})(990) \, CV}{9.110 \times 10^{-31} kg}}$$

$$= \sqrt{3.482 \times 10^{14} \frac{JV^{-1}V}{kg}}$$

$$= \sqrt{3.482 \times 10^{14} \frac{kg \, m^2 \, sec^{-2}}{kg}}$$

$$= \sqrt{3.482 \quad 10^{14}\,\mathrm{m^2\,s^{-2}}}$$

$\therefore$ $v = 1.866 \times 10^{7}\,\mathrm{m\,s^{-1}}$

Similarly, for a potential difference of 1.01 kV,

$$v = 1.885 \times 10^{7}\,\mathrm{m\,s^{-1}}$$

Hence $\delta v = (1.885 - 1.866) \times 10^{7}\,\mathrm{m\,s^{-1}}$

$$= 0.019 \times 10^{7}\,\mathrm{m\,s^{-1}}$$

$$= 1.9 \times 10^{5}\,\mathrm{m\,s^{-1}}$$

Solving for δp in equation (2) gives

$$\delta p = (9.110 \times 10^{-31}\,\mathrm{kg})\,(1.9 \times 10^{5}\,\mathrm{m\,s^{-1}})$$

$$= 17.309 \times 10^{-26}\,\mathrm{kg\,m\,s^{-1}}$$

$$= 1.7 \times 10^{-25}\,\mathrm{kg\,m\,s^{-1}}.$$

Rearranging equation (1) to solve for δz gives

$$\delta z \geq \left[\frac{\hbar}{2}\right] \Big/ \delta p = \frac{1.054 \times 10^{-34}\,\mathrm{Js}}{2(1.7 \times 10^{-25}\,\mathrm{kg\,m\,s^{-1}})}$$

$$= 3.09 \times 10^{-10}\,\mathrm{Js / kg\,m\,s^{-1}}$$

$$= 3.09 \times 10^{-10}(\mathrm{kg\,m^2 / s^2})\mathrm{s / kg\,m\,s^{-1}}$$

or $\delta z = 3.09 \times 10^{-10}\,\mathrm{m} = 309\,\mathrm{pm} = 3.09$ Å.

11

Macromolecules

Q.1 : For a solution of cellulose trinitrate (M = 140,000) in acetone, $dn/dc = 0.105\ cm^3.g^{-1}$ and $n_o = 1.3589$. Calculate the ratio of intensities of transmitted to incident light at wavelenths of 400 and 700 nm through 1.00 cm thickness of a solution of the polymer containing 2.00 g per 100 cm^3

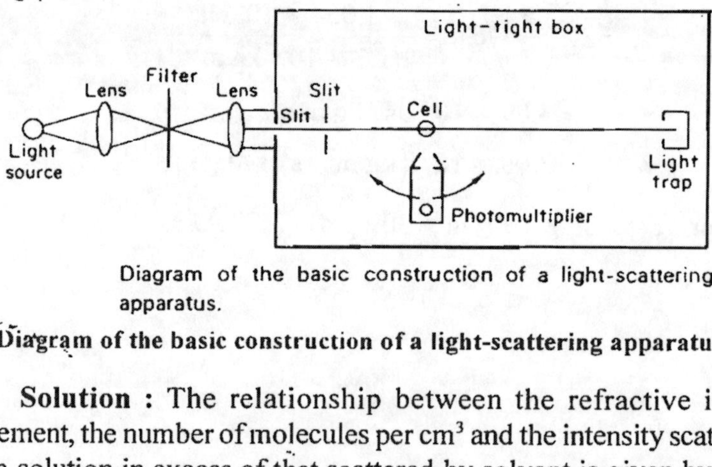

Diagram of the basic construction of a light-scattering apparatus.

Diagram of the basic construction of a light-scattering apparatus.

Solution : The relationship between the refractive index increment, the number of molecules per cm^3 and the intensity scattered from solution in excess of that scattered by solvent is given by

$$\frac{i}{I_0} = \frac{2\pi^2 n_0^2 (1+\cos^2\theta)(dn/dc)^2 Mc}{N\lambda^4 r^2}$$

where i = intensity

I_o = intensity of incident unpolarized light

n_o = refractive index of solvent

$\frac{cN}{M}$ = number of molecules per cm^3

$$\frac{dn}{dc} = \text{refractive index increment}$$

r = distance from the scattering molecule

θ = angle measured from the emergent beam

λ = wave length.

If the above expression is integrated over all angles to obtain the total scattered intensity, it is found that the transmitted intensity is given by

$$I = I_0 e^{-\tau x} \tag{1}$$

where I = transmitted intensity

τ = turbidity

x = thickness of the cell in the direction of the incident beam.

Observe that
$$\tau = \frac{32\pi^2 n_0^2 (dn/dc)^2 \; Mc}{3N\lambda^4}$$

and x = 1 cm thick.

Equation (1) can be rearranged to give

$$I/I_0 = e^{-\tau x} \tag{2}$$

where I/I_0 = ratio of intensities of transmitted to incident light.

Solving for τx gives

for λ = 400 cm,
$$\tau x = \frac{(3130 \times 10^{-23})(1 \text{ cm})}{256 \times 10^{-20} \text{ cm}^4}$$

$$= 0.0122$$

Therefore,
$$I/I_0 = e^{-\tau x} = e^{-0.0122}$$
$$= 0.988.$$

For λ = 700,
$$\tau x = \frac{(3130 \times 10^{-23})(1 \text{ cm})}{2401 \times 10^{-20} \text{ cm}^4}$$
$$= 0.\ 0013$$

Therefore, $I/I_0 = e^{-\tau x} = e^{-0.013} = 0.999.$

Q. 2 : (a) Briefly discuss the limitation of end-group analysis for molecular weight determination.

(b) Fifty grams of a polyester was separated into homogeneous molecular weight fractions by sedimentation. The carboxylic end-groups in each fraction were titrated by base with the following results :

Weight of fraction (grams) :	1	10	25	9	5
Moles of base titrant :	10^{-3}	10^{-4}	10^{-5}	10^{-6}	10^{-7}

Determine the weight average and the number average molecular weights of the 50 gram sample.

Solution : (a) The determination of molecular weight of a macromolecule (or polymer) is done by numerous methods of which one of them is End-group analysis. Amongst others are light scattering, colligative properties, etc.

The End-group analysis is the limitation of long chains to end groups of macromolecules (or polymers). The functional group analysis requires that a known amount of determinable groups per molecule be present, for example, the determination of amino groups in polyesters and polyamides. the limitation of end-group analysis is that chemical methods for molecular weight determination become insensitive at molecular weights higher than about 25000, as the fraction of end groups become too small to be detected with precision. Therefore, the viscosity is increased as well as the effect of crosslinking.

(b) The weight average molecular weight is given by

$$\overline{M}_W = \frac{\sum W_i M_i}{\sum W_i} = \frac{\sum N_i M_i^2}{\sum N_i M_i} = \sum W_i M_i \qquad (1)$$

where $\overline{M}_W$ = weight average molecular weight

M_i = molecular weight of the ith species;

W_i = weight of material with molecular weight M_i;

N_i = number of moles with molecular weight M_i;

W_i = weight fraction of molecules of the ith species.

From the given data, total weight = 50 grams. Therefore,

$$W_1 = \frac{1}{50},\ W_2 = \frac{10}{50},\ W_3 = \frac{25}{50},\ W_4 = \frac{9}{50} \text{ and } W_5 = \frac{5}{50}.$$

The corresponding molecular weights using the relation that molecular weight = weight/moles are

$$M_1 = \frac{1}{10^{-3}} = 10^3,\ M_2 = \frac{10}{10^{-4}} = 10^5,\ M_3 = \frac{25}{10^{-5}} = 25 \times 10^5,$$

$$M_4 = \frac{9}{10^{-6}} = 9 \times 10^6 \text{ and } M_5 = \frac{5}{10^{-7}} = 5 \times 10^7.$$

Inserting these values into equation (1) gives

$$\overline{M}_W = \sum W_i M_i = \frac{1}{50} = 10^3 + \frac{10}{50} \times 10^5 + \frac{25}{50} \times 25 \times 10^5$$

$$+ \frac{9}{50} \times 9 \times 10^6 + \frac{5}{50} \times 5 \times 10^7$$

$$= 20 + 2000 + 1250000 + 1620000 + 5000000$$

Therefore, $\overline{M}_W = 7890020.$

The number average molecular weight is defined as

$$\overline{M}_n = \frac{1}{\sum (w_i / M_i)}$$

$$= \frac{1}{\frac{0.02}{10^3} + \frac{0.2}{10^5} + \frac{0.5}{2.5 \times 10^6} + \frac{0.18}{9 \times 10^6} + \frac{0.1}{5 \times 10^7}}$$

$$= \frac{1}{0.00002 + 0.000002 + 0.0000002 + 0.00000002 + \frac{.1}{5 \times 10^7}}$$

Therefore, $\overline{M}_n = \frac{1}{0.000022222}$

or $\overline{M}_n = 4504$

Q. 3 : Physical Constants of Proteins at 20° in water

Protein	$S \times 10^{13}$, s	$D \times 10^{11}$ $m^2\ s^{-1}$	$\overline{v}$, $cm^3\ g^{-1}$
Beef insulin	1.7	15	0.72
Lactalbumin	1.9	10.6	0.75
Ovalbumin	3.6	7.8	0.75
Serum albumin	4.3	6.15	0.735
Serum globulin	7.1	4.0	0.75
Urease	18.6	3.4	0.73
Tobacco mosaic virus	185	0.53	0.72

The physical constants of proteins at 20° in water are given in the table above. The table tabulates values of the sedimentation coefficient s, the diffusion coefficient D, and the partial specific volume $\overline{v}$, for various proteins. Using these data estimate the maximum molecular weight of ovalbumin. State any assumption(s) made.

Solution : Assuming that ovalbumin is spherical, the molecular weight may be estimated from diffusion coefficient. This diffusion coefficient is related to frictional coefficient f, by the expression

$$D = \frac{RT}{Nf} \tag{1}$$

where D = diffusion coefficient

R = universal gas constant

T = temperature

N = Avogadro's number

f = frictional coefficient

Equation (1) was first obtained by Einstein.

For spheres and nonturbulent flow, the frictional coefficient is given by

$$f = 6\pi\eta r \tag{2}$$

where η = coefficient of viscosity = 001005 Js/m³

r = radius of the sphere.

Combining equations (1) and (2) yields

$$D = \frac{RT}{N6\pi\eta r}. \tag{3}$$

Since the estimation of molecular weight of the particle is needed, the expression

$$\frac{M\bar{v}}{N} = \frac{4}{3}\pi r^3 \tag{4}$$

is used,

where M = molecular weight

$\bar{v}$ = partial specific volume.

Solving for r in equation (4) gives

$$r^3 = \frac{3M\bar{v}}{4\pi N}$$

$$r = \left(\frac{3M\bar{v}}{4\pi N}\right)^{\frac{1}{3}} \tag{5}$$

Substituting this expresson for r in equation (3) gives

$$D = \frac{RT}{46\pi\eta\left(\frac{3M\bar{v}}{4\pi N}\right)^{\frac{1}{3}}}$$

$$= \frac{RT}{46\pi\eta}\left(\frac{4\pi N}{3M\bar{v}}\right)^{\frac{1}{3}} \tag{6}$$

Solving equation (6) for M with the given data yields

$$7.8 \times 10^{-11}\,m^2\,s^{-1} = \left[\frac{(8.31\,J\,K^{-1}\,mol^{-1})\,(293\,K)}{(6.02 \times 10^{23}\,mol^{-1})\,6\pi\,(0.001005 J\,m^{-3}\,s)}\right]$$

$$\times \left[\frac{4\pi\,(6.02 \times 10^{23}\,mol^{-1})}{3M\,(0.75 \times 10^{-3}\,m^3\,kg^{-1})}\right]^{\frac{1}{3}}$$

and $M = 69.0$ kg mol^{-1}

$= 69000$ g mol^{-1}

Q. 4 : The density of water at 20° is 9.9982 g cm^{-3}. Using the data in the table given below, calculate the molecular weight of serum albumin.

Physical Constants of Proteins at 20° in water

Protein	$S \times 10^{13}$, s	$D \times 10^{11}$, $m^2 s^{-1}$	$\bar{v}$, $cm^3 g^{-1}$
Beef insulin	1.7	15	0.72
Lactalbumin	1.9	10.6	0.75
Ovalbumin	3.6	7.8	0.75
Serum albumin	4.3	6.15	0.735
Serum globulin	7.1	4.0	0.75
Urease	18.6	3.4	0.73
Tobacco mosaic virus	185	0.53	0.72

Solution : The equation on which this calculation is based may be derived by setting the force of the centrifugal field on the particle equal to the frictional force, f (dr/dt), where f is the frictional coefficient of the molecule and dr/dt is the speed of sedimentation. The force of the field on a particle of mass m and partial specific volume $\bar{v}$ suspended in a medium of density ρ is

$$m(1-\bar{v}\rho)\,\omega^2 r = \frac{M}{N}(1-\bar{v}\rho)\omega^2 r \tag{1}$$

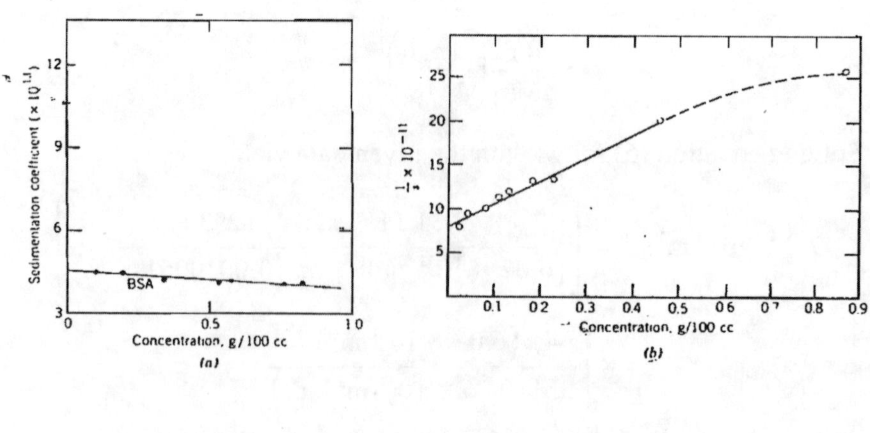

(a) The effect of concentration on the sedimentation coefficient of bovine serum albumin (BSA; Kegeles and Gutter[60]; Kahler[62]).

(b) Kahler's data plotted as 1/s versus c.

where $(1-\bar{v}\rho)$ = buoyancy factor

M = molecular weight

$\omega^2 r$ = centrifugal acceleration

N = Avogadro's number.

The sedimenting molecule or particle will be accelerated by the field until its velocity is such that the frictional force is equal to the force of the field. That is

$$f\frac{dr}{dt} = \frac{M}{N}(1-\bar{v}\rho)\omega^2 r$$

or
$$\frac{dr/dt}{\omega^2 r} = \frac{M}{N}\frac{(1-\bar{v}\rho)}{f} \quad (2)$$

By definition, $\frac{dr/dt}{\omega^2 r} = S$, where S = sedimentation coefficient. Using the relation obtained by Einstein,

$$D = \frac{RT}{Nf},$$

or
$$f = \frac{RT}{DN} \quad (3)$$

where f = frictional coefficient

D = diffusion coefficient

R = universal gas constant

T = absolute temperature

the combination of equations (2) and (3) will yield

$$S = \frac{M}{N}\frac{(1-\bar{v}_0)}{(RT/DN)}$$

$$= \frac{M}{N}(1-\bar{v}_0)\frac{DN}{RT}$$

$$= \frac{M(1-\bar{v}_0)D}{RT}$$

Solving for M gives

$$M = \frac{SRT}{D(1 - \bar{v}_0)}.$$

Using the table given above,

$$M = \frac{(4.3 \times 10^{-13}\,s)(8.31\,J\,K^{-1}\,mol^{-1})(293\,K)}{(6.15\times10^{-11}\,m^2\,s^{-1})\,1-(0.735\times10^{-3}\,m^3\,kg^{-1})(0.9982\times10^3\,kg\,m^{-3})}$$

$= 64.0 \text{ kg mol}^{-1}$

$= 64000 \text{ kg mol}^{-1}$

Q. 5 : For horse hemoglobin in water solution at 20°C, $D = 6.3 \times 10^{-7}$ cm². s⁻¹, $s = 4.41 \times 10^{-13}$ s, $\bar{v} = 0.749$ cm³ g⁻¹, $\rho = 0.9982$ g.cm⁻³. Calculate the molecular weight.

Solution : The hemoglobin sedimenting particle will be accelerated by the field until its velocity is such that the frictional force is equal to the force of the field. That is

$$f\frac{dr}{dt} = \frac{M}{N}(1 - \bar{v}\rho)\omega^2 r$$

or

$$\frac{dr/dt}{\omega^2 r} = \frac{M(1-\bar{v}\rho)}{Nf} \qquad (1)$$

where
M = molecular weight

N = Avogadro's number

f = frictional coefficient

$(1 - \bar{v}\rho)$ = buoyancy factor

dr/dt = speed of sedimentation

$\omega^2 r$ = centrifugal acceleration.

The speed of sedimentation divided by the centrifugal acceleration is called the sedimentation coefficient S. Therefore, equation (1) becomes

$$S = \frac{M\left(1 - \bar{v}\rho\right)}{Nf} \tag{2}$$

The diffusion coefficient D, is defined as

$$D = \frac{RT}{Nf}$$

Solving for f gives,

$$f = \frac{RT}{DN} \tag{3}$$

Substitute this expression for f into equation (2) to obtain

$$S = \frac{M(1 - v\rho)}{N\left(RT / DN\right)}$$

Solving for M gives

$$M = \frac{RTS}{D\left(1 - \bar{v}\rho\right)} \tag{4}$$

Since all the variables in equation (4) are known, the molecular weight of the horse hemoglobin in water solution is

$$M = \frac{\left(8.31 \times 10^{7}\,\text{erg K}^{-1}\,\text{mol}^{-1}\right)\left(293\text{ K}\right)\left(4.41 \times 10^{-13}\text{ sec}\right)}{\left(6.3 \times 10^{-7}\text{cm}^{2}\text{ sec}^{-1}\right)\left[1 - \left(0.749\text{ cm}^{3}\text{ g}^{-1}\right)\left(0.9982\text{ cm}^{-3}\text{ g}\right)\right]}$$

$$= \frac{17043.81}{1 - 0.7477}\text{ g mol}^{-1} = \frac{17043.81}{0.2523}\text{ g mol}^{-1}$$

$$= 67553.745\text{ g mol}^{-1} \cong 67554\text{ g mol}^{-1}.$$

Q. 6 : The relative viscosity of a polymer solution containing 1.00g of polymer in 100 cm^3 is 2.800. A solution half as concentrated has a relative viscosity of 1.800.

(a) Calculate the intrinsic viscosity. (Assume that a graphical treatment would give a straight line, and calculate the intercept analytically).

(b) If the appropriate values of K and a in the Mark-Houwink equation are 5.00×10^{-4} and 0.600, calculate the molecular weight of the polymer.

Solution : (a) Viscosity measurements provide an additional method of molecular weight estimation. It was determined by Einstein that for a dilute suspension of spheres in a viscous medium

$$\lim_{\varphi \to 0} \frac{(\eta / \eta_0) - 1}{\varphi} = 2.5 \qquad (1)$$

where η and η_0 are the viscosities of the solution or suspension and pure solvent, respectively, and φ is the volume fraction of the solution occupied by the spheres. The fraction η/η_0 is the relative viscosity η_r, and $[(\eta/\eta_0) - 1]$ is called the specific viscosity η_{sp}. For polymer solutions the concentration c is usually given as grams per cubic centimeter and $\eta_{sp}/100c$ is called the reduced viscosity.

The limiting value of the reduced viscosity as c approaches zero is the intrinsic viscocity $[\eta]$.

By definition,

$$\frac{1}{c}(\eta_r - 1) = kc + [\eta] \qquad (2)$$

where $\qquad c$ = concentration

$$k = \frac{1/c_1(\eta_{r_1} - 1) - 1/c_2(\eta_{r_2} - 1)}{c_1 - c_2}$$

Since c_1 is given to be 1.00 g dl^{-1} and $c_2 = 0.5$g dl^{-1}.

$$\frac{1}{c_1}(\eta_{r_1} - 1) = \frac{1}{1}(2.800 - 1) = 1.800$$

and $$\frac{1}{c_2}(\eta_{r_2} - 1) = \frac{1}{0.5}(1.800 - 1) = 1.600$$

Now, solving for k gives

$$k = \frac{1.800 - 1.600}{1.0 - 0.5}$$

$$k = 0.400 \text{ g}^{-2} \text{ dl}^2$$

Since equation (2) is linear,

$$[\eta] = \frac{1}{c_1}(\eta_{r_1} - 1) - kc_1 \qquad (3)$$

and $$[\eta] = \frac{1}{c_2}(\eta_{r_2} - 1) - kc_2. \qquad (4)$$

From equation (3)

$$[\eta] = 1.800 - 0.400\,(1.0)$$

$$= 1.400 \text{ g}^{-1} \text{ dl},$$

and equation (4) gives

$$[\eta] = 1.600 - 0.400\,(0.5)$$

$$= 1.400 \text{ g}^{-1} \text{ dl}.$$

(b) The dependence of the intrinsic viscosity on the molecular weight of a series of samples of the same polymer in a given solvent and at a constant temperature is given by equation (2) written in the form

$$[\eta] = KM^a \qquad (5)$$

The exponant a can vary depending on the stiffness of the polymer and together with K they are treated as constants that may be determined by measuring the intrinsic viscosities of a series of polymer samples for which molecular weights have been measured by another method; i.e., light scattering.

Taking the log of both sides of equation (5) gives

$$\log[\eta] = \log(KM^a)$$

$$= \log K + \log M^a$$

$$\log[\eta] = \log K + a \log M \qquad (6)$$

Rearranging eqaution (6) to solve for log M gives

$$\log [\eta] = \log K + a \log M$$

or $$a \log M = \log [\eta] / K$$

$$\log M = \frac{(\log [\eta] / K)}{a}$$

$$= \left(\log \frac{1.400}{5.00 \times 10^{-4}} \right) \Big/ 0.600$$

$$= \frac{3.447}{0.600} = 5.745$$

$$\therefore \qquad M = 5.56 \times 10^5 \text{ g mol}^{-1}.$$

Q. 7 : Given that $\overline{X}_n = 100$ for each of the following macromolecules: polystyrene, polycaprolactam and nylon (66), otherwise known as poly (hexamethylene adipamide), compute $\overline{M}_n$ for each of them.

Solution : The number average molecular weight of a polymer is the product of the molecular weight of the monomer that make up the polymer and the number average degree of polymerization. This is mathematically defined as

$$\overline{M}_n = m\overline{X}_n,$$

where $\overline{M}_n$ = number average molecular weight

m = molecular weight of the monomer

$\overline{X}_n$ = number average degree of polymerization.

If more than one monomer make up the polymer, then m = mean of the molecular weights.

For polystyrene, the monomer is represented as

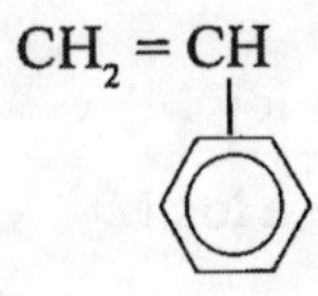

Therefore, m = 12 + 2 + 12 + 6 (12) + 5 (1) + 1 = 104.

Consequently, $\overline{M}_n = m\overline{X}_n$

$= (104)\,(100)$

$= 10400$

For polycaprolactam, the monomer is represented as

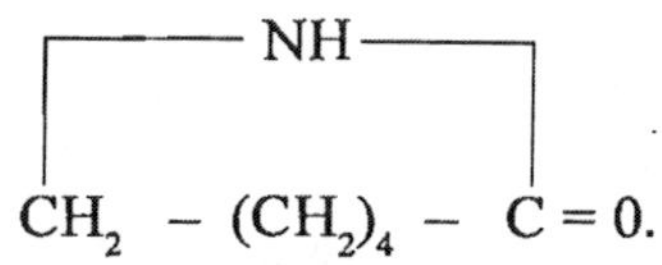

Therefore, $m = 12 + 2 + 4\,(12) + 4\,(2) + 12 + 14 + 1 + 16 = 113$

Consequently, $\overline{M}_n = m\overline{X}_n$

$= (113)\,(100)$

$= 11300.$

Finally, the monomers for nylon (66) are represented as $NH_2 - (CH_2)_6 - NH_2$ and $HOOC - (CH_2)_4 - COOH$. Their molecular weights are

$$14 + 2 + 6\,(12) + 6\,(2) + 14 + 2 = 116$$

and $$1 + 16 + 16 + 12 + 4\,(12) + 4\,(2) + 12 + 16 + 16 + 1$$

$$= 146 \text{ respectively.}$$

Therefore, $m = \text{mean molecular weight} = \dfrac{116 + 146}{2} = 13$

Observe that to make a reccurence unit, one H_2O per monomer is eleminated. As a result, effective $m = 131 - 18 = 113$.

Consequently, $\overline{M}_n = m\overline{X}_n$

$= (113)\,(100)$

$= 11300.$

Q. 8 : Calculate the most probable value of the chain length of normal $C_{20}H_{42}$ give the C – C bond length of 0.15 nm and the bond angle of 109° 28'.

Solution : If a molecule has the typical zig-zag shape of a carbon backbone, then the length of a unit link of the backbone can be found using the following equation.

$$a = \ell\left(\frac{1 + \cos\alpha}{1 - \cos\alpha}\right)^{1/2}$$

where a = length of a unit link of the backbone

ℓ = C–C bond length

α = bond angle.

To put 28' into decimal notation, divide 28' by 60'/ degree. Hence, 109°28' is now equal to 109.467°.
Therefore,

$$a = 0.15 \text{ nm}\left(\frac{1 + \cos 109.467^\circ}{1 - \cos 109.467^\circ}\right)^{1/2}$$

$$= 0.15 \text{ nm}\left(\frac{1 + -0.333}{1 - -0.333}\right)^{1/2}$$

$$= 0.106 \text{ nm}$$

For $C_{20}H_{42}$ there are twenty carbon atoms in the chain but there are nineteen bonds. The most probable value of the chain length is given by

$$r = \beta^{-1} = a\left(\frac{2N}{3}\right)^{1/2}$$

where N = number of links in the backbone.
Therefore,

$$r = 0.106 \text{ nm}\left(\frac{2(19)}{3}\right)^{1/2} = 0.377 \text{ nm.}$$

Q. 9 : The following data are available for the solutions of γ globulin in 0.15M NaCl at 37°C

c, concentration of γ globulin, g per 100 ml	π, osmotic pressure, nm H_2O
19.27	453
12.35	253
5.81	112

Calculate the molecular weight of the γ globulin.

Solution : The osmotic pressure of non-electrolyte can be represented by a power series in the concentrations. That is,

$$\pi = (RT/M)c + BC^2 + BC^3 + ... \qquad (1)$$

where c = the mass concentration. In dilute solutions, equation (1) can be truncated to give

$$\frac{\pi}{c} = \frac{RT}{M} + BC \qquad (2)$$

The value of B for particular solvents and temperatures may be set equal to zero. Therefore, equation (2) becomes

$$\frac{\pi}{c} = \frac{RT}{M} \qquad (3)$$

A plot of π/c vs c should give a straight line in the region of low concentration. The intercept at c → 0 yields RT/M, from which the molecular weight is determined.

For the first solution

$$\frac{\pi}{c} = \frac{453 \text{ mm } H_2O}{19.27 \text{ g dl}^{-1}}$$

$$= 23.5 \text{ mm } H_2O \text{ dl g}^{-1}.$$

Similarly, for the other two solutions,

$$\frac{\pi}{c} = 20.5 \text{ and } 19.3 \text{ mm } H_2O \text{ dl g}^{-1}.$$

From the plot of π/c vs c, the intercept at c → 0 gives 18.6 mm H_2O dl g^{-1}

as $$\lim_{c \to 0} \left(\frac{\pi}{c}\right) = 18.6 \text{ mm } H_2O \text{ dl g}^{-1}$$

$$= \frac{\left(18.6 \text{ mm H}_2\text{O dl g}^{-1}\right)\left(0.100 \text{ liter dl}^{-1}\right)}{\left(13.56 \text{ mm H}_2\text{O torr}^{-1}\right)\left(760. \text{ torr atm}^{-1}\right)}$$

$$= 1.804 \times 10^{-4} \text{ liter atm g}^{-1}$$

and $$M = RT/\lim(\pi/c)$$

$$= \frac{\left(0.08206 \text{ liter atm }^\circ\text{K}^{-1}\text{ mole}^{-1}\right)\left(310\ ^\circ\text{K}\right)}{1.804 \times 10^{-4} \text{ liter atm g}^{-1}}$$

$$M = 1.41 \times 10^{5} \text{ g mole}^{-1}.$$

12

Phase Rule

Q. 1 : The diagram below is an example of a phase diagram for a pure substance. To what phases do the regions A, B, and C correspond ?

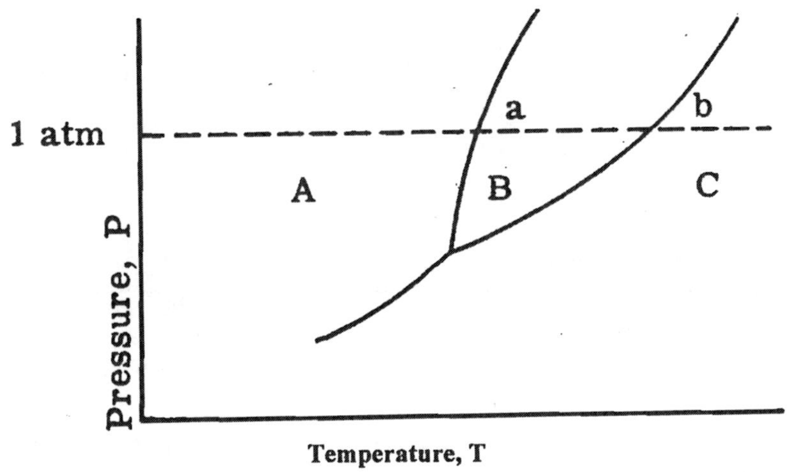

Solution : Following the 1 atm constant pressure line from left to right, we are proceeding from low values of the temperature to high values. Therefore, we will intersect the three regions in the order solid-liquid-vapour. The regions A, B, C hence correspond to the solid, liquid and vapour phases, respectively. Point a denotes the normal freezing (melting) point of the substance and the point b denotes the normal boiling point.

Q. 2 : Draw a labelled phase diagram for a substance z which has the following properties; normal boiling point = 220°C, normal freezing point 80°C, and triple point 60°C and .20 atm. Predict the freezing and boiling, if the pressure were .80 atm ?

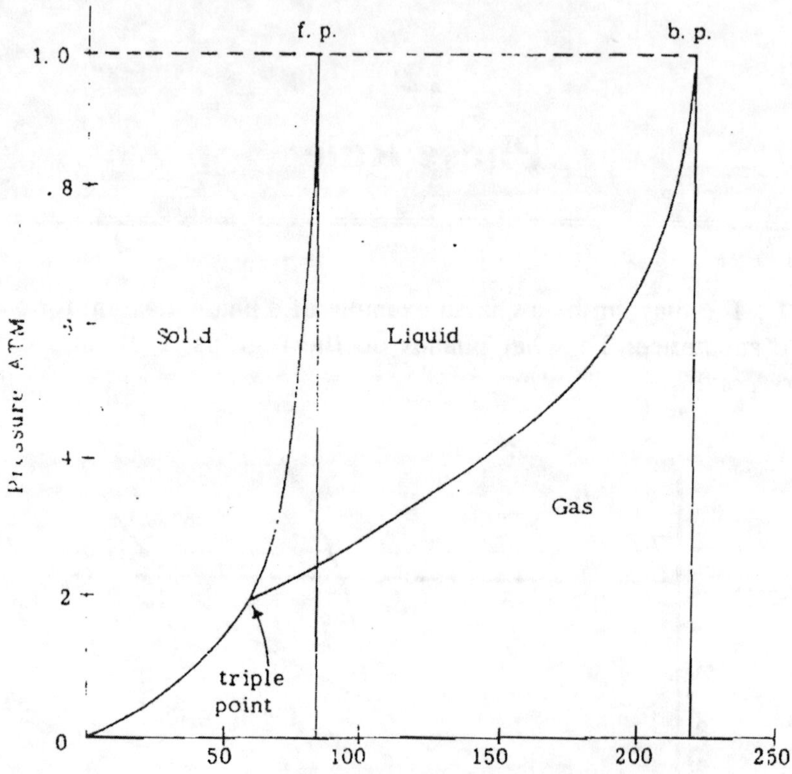

Solution : To draw this diagram you want to understand all the terms involved. The relation between solid, liquid, and gaseous states as a function of the given temperature and pressure can be summarized on a graph known as a phase diagram. From the given experimental observations, you can draw the diagram. The lines which separate the states in a diagram represent an equilibrium between the phases. The intersection of the three lines is called the tripple point, where all three phases are in equilibrium with each other. By normal boiling and melting points you mean those readings taken at 1 atm. Thus, the phase diagram can be written as shown in the accompanying figure.

From the diagram you see that if the atm was .80, the b.p. and f.p. would drop, respectively, to 215°C and 85° C.

Q. 3 : The vapour-liquid equilibrium data of the system chloroform-ethanol at 45°C are listed in table 1.

x_1	y_1	p (mmHg)	p_1 (mmHg)	p_2 (mmHg)
0.1260	0.3974	249.92	99.30	150.62
0.2569	0.6060	329.62	199.60	130.02
0.4015	0.7143	391.51	279.50	112.01
0.6283	0.7954	438.89	349.00	89.89
0.8206	0.8516	455.56	388.60	66.96
0.9557	0.9319	448.49	418.00	30.49

Table 1

Assuming ideal behaviour of the vapour phase, prove the thermodynamic consistency of the experimental data.

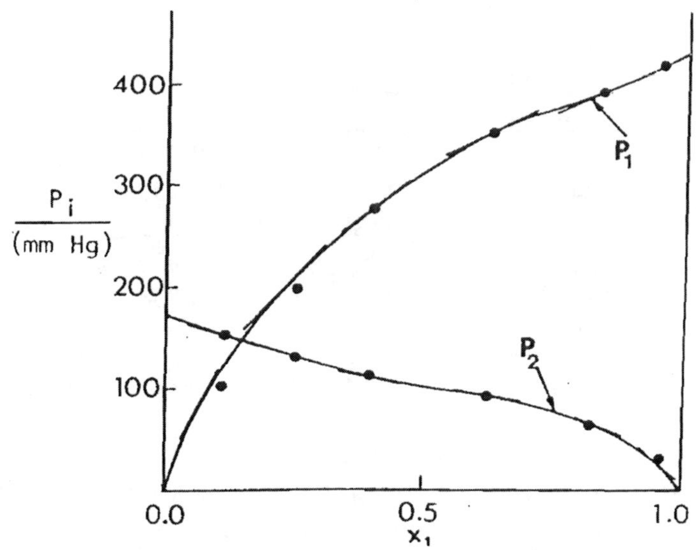

Fig. 1. Variation of the partial pressure with composition : system, chloroformethanol at 45°C.

Solution : The Gibbs-Duhem equation applies to this problem. When the vapour phase is assumed to be ideal (that is $f_i = p_i$), the

Gibbs-Duhem equation can be written in the form

$$\Delta = \frac{x_1}{P_1}\frac{dP_1}{dx_1} - \frac{x_2}{P_2}\frac{dP_2}{dx_2}$$

where P_i = partial pressure of component i
f_i = fugacity of component i
x_i = mole fraction of component i
Δ = deviation (if any)

Note that the smaller Δ is the more consistent of experimental data. To prove the consistency of this data, the slopes $\frac{dP_1}{dx_1}$, $\frac{dP_2}{dx_2}$, P_1 and P_2, at the corresponding mole fractions must be known. The desired information can be obtained by plotting P_1 and P_2 against the mole fractions as shown below in Fig. 1. The slopes are computed for the given values of x_1. Then the Δ values can be computed. The respective values are tabulated in table 1.

Table 1

$\frac{dP_1}{dx_1}$	$\frac{dP_2}{dx_2}$	Δ
$\frac{\text{mmHg}}{\text{mole fraction}}$		
792	– 176	– 0.016
724	– 140	+ 0.132
416	– 118	– 0.033
229	– 96	+ 0.015
176	– 153	– 0.038
290	– 472	– 0.023

Since the Δ values are small, the data is thermodynamically consistent.

Q. 4 : Fig. 1 shows the boiling point diagram for two miscible liquids. (a) Show the composition of the first and last vapour, if 50 mole % solution is boiled in an open container. (b) Find the liquid composition when half the solution is vapourized during heating in a closed system at 1 atm constant pressure. What is the temperature ?

Solution : (a) The composition of the liquid and vapour phases in equilibrium with each other are given by the ends of the horizontal lines called tie lines. One such line is line L of Fig. 2. The composition of the first vapor corresponding to 50 mole % solution is obtained using the top curve at about $N_B = 0.1$ at a temperature of about 68°C. The last vapor composition will correspond to the maximum boiling point at T_{bmax}. Thus $N_B = 0.95$.

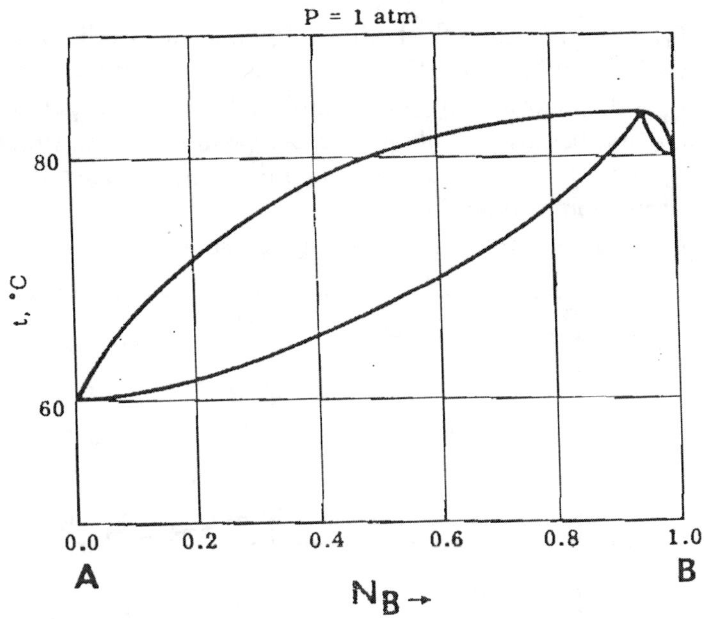

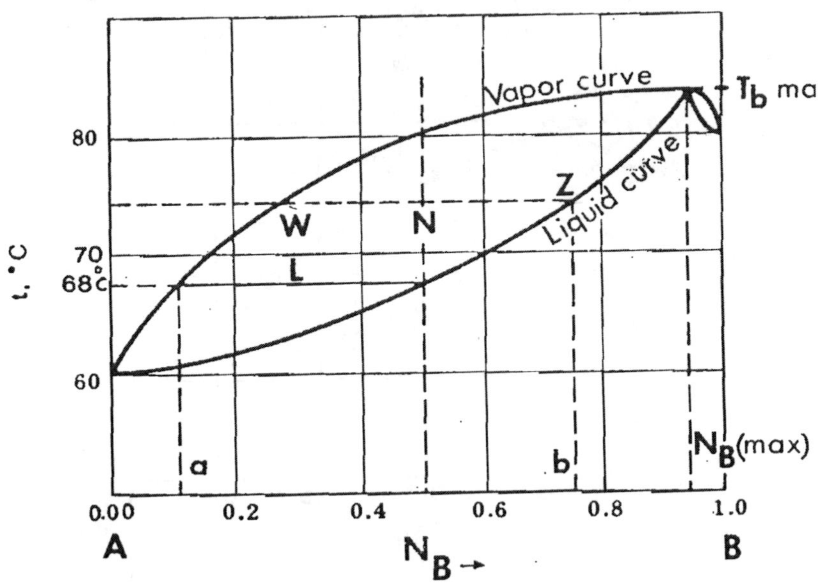

(b) The temperature corresponding to half vaporization in a closed system will be such that the system composition of 50 mole % bisects the tie line at point N. This means that distance WN = distance NZ. This occurs at about 75°C and the liquid composition corresponding to this point is $N_B = 0.75$.

Q. 5 : Fig. 1 shows the boiling point plot for two liquids A and B. If a solution of mole fraction $N_B = 0.4$ is boiled openly in a beaker until $y_B = 0.4$, show on the graph in Fig. 1 (a) the temperature the solution first begins to boil; (b) the composition of the first vapor coming off; (c) the boiling point when the vapor reaches $Y_B = 0.4$; (d) composition of final liquid.

In all cases give approximate numerical answers.

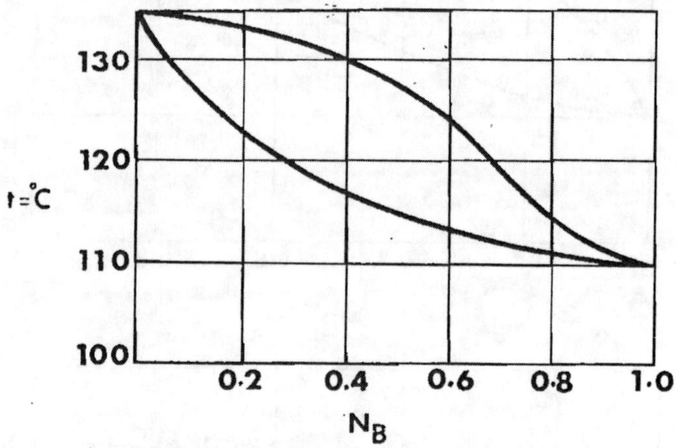

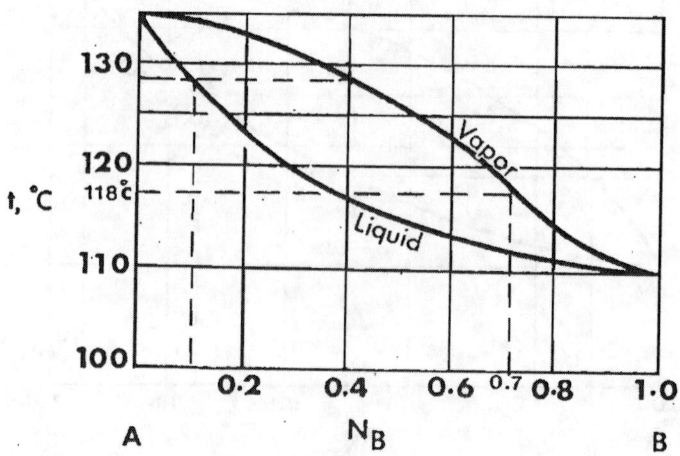

Solution : (a) The liquid begins to boil at a temperature given by the lower liquid curve. Thus at a composition of $N_B = 0.4$ the corresponding temperature using the lower curve is about 118°C.

(b) The composition of the first vapor coming off is found using the top curve at $N_B = 0.4$. ∴ $Y_B = 0.7$.

(c) When the vapor reaches $Y_B = 0.4$, using the top curve at $N_B = 0.4$, the boiling point is about 126°C.

(d) The composition of the final liquid is found using the lower curve at the boiling point of 126°C and it is estimated to be about $N_B = 0.12$. These are all shown in Fig. 2.

Q. 6 : Figure 1 is a diagram for two substances that show partial miscibility with each other in the solid state. This type of diagram results from the ability of one substance to penetrate the empty spaces, etc., of the lattice of the other. There are limitations on this type of solubility and the maximum is indicated by the curved lines enclosing areas # 1 and # 4. Identify the phases present in the numbered areas.

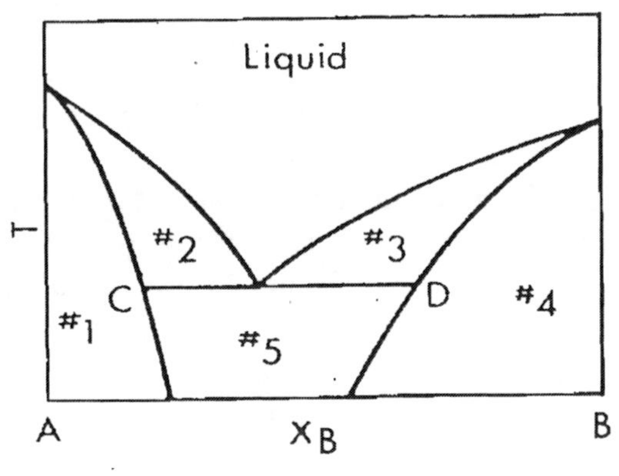

Solutions : This is a T vs X_B plot where T = temperature and X_B = composition of component B. The points A and B represent pure A and pure B respectively. The horizontal line between points C and D is called a tie line. The composition of the liquid and solid phases in equilibrium with each other is given by the ends of the tie line. Thus the liquid and solid solution of component B in saturated component A are in equilibrium in area # 2. Also the liquid and solid solution of component A in saturated component B are in equilibrium in area # 3. In area # 5, solid solution of B in saturated A and solid solution of A in saturated B are in equilibrium. Areas #1 amd #4 are single phase regions. They consist of unsaturated solid solution of B in A and of A in B respectively.

Q. 7 : Identify the areas labelled A, B, C in Figure 1 of a pure substance. (P denotes vapor pressure except for the line between areas A and B for which it denotes external pressure.)

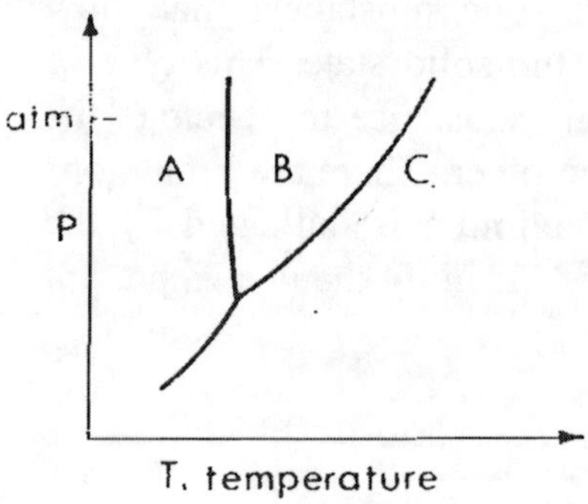

Fig. 1.

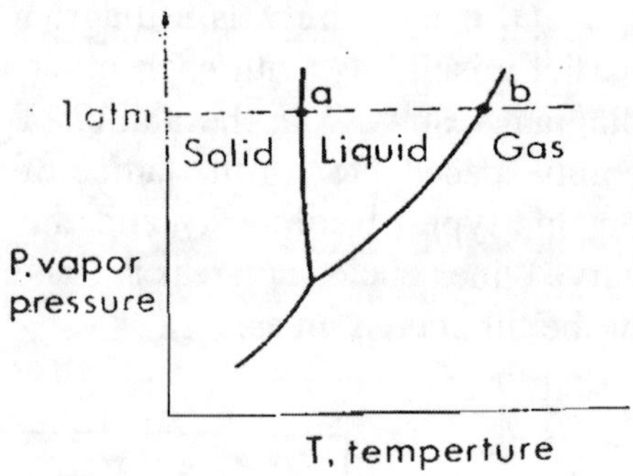

Fig. 2.

Solution : If one maintains a constant pressure, say 1 atm, as indicated in Figure 2 and starting at a low temperature to a high temperature, the dashed line of Figure 2 is traced out. Because one is moving from a low to a high temperature, it is logical to say that areas A, B, C corresponds to solid phase, liquid phase and gas phase respectively. Point a indicates the freezing point of the substance and point b, the boiling point.

Q. 8 : Consider the hypothetical phase diagram shown in Fig. 1. In such a diagram certain areas will be one-phase areas and others will be two-phase areas. The compositions of the phases in equilibrium in the two-phase areas will be determined by horizontal "tie lines." What would be the masses of the phases for a system containing

0.050 kg of A and 0.050 kg of B in equilibrium if $(\text{wt\% B})_1$ =
and $(\text{wt \% B})_2$ = 85.5% ?

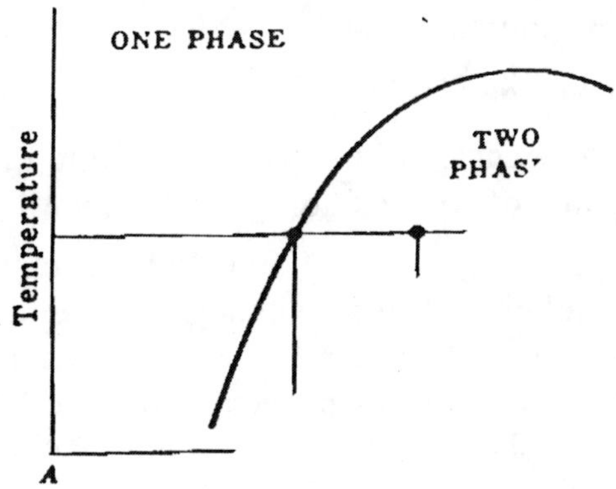

Solut'

$$m_1 + m_2 = 0.100$$

$$\frac{m_1}{m_2} = 1.775$$

Upon solving them simultaneously,

$$m_1 = 0.064 \text{ kg and } m_2 = 0.036 \text{ kg.}$$

Q. 9 : Describe the changes in the system shown in Fig. 1 as salt B is added to a solution of composition given by point 1.

Solution : The line between Point 1 and Vertex B represents addition of pure B to the solution of composition given by Point 1. As we add B to the solution, the system gets richer in B and less rich in A and C. If enough B is added for the bulk concentration to reach Point 2, solid B becomes a second phase in equilibrium with a solution phase of composition given by Point 2. If we add more B, the composition does not change but the relative amounts do, indicating saturation.

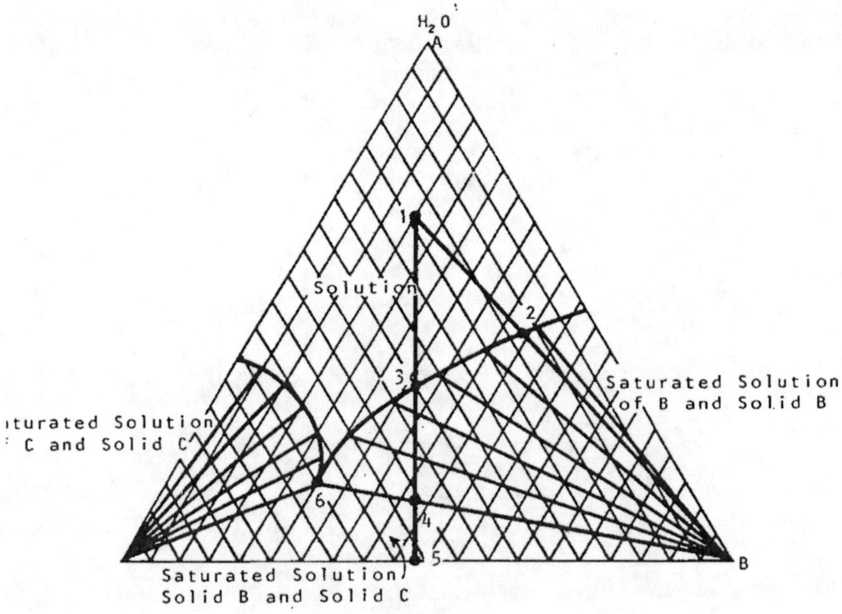

Fig. 1

13

Ionic Equilibria and Electrochemistry

Q. 1 : A moving boundary experiment was performed at 25°C with 0.100 M KCl on the bottom and (initially) 0.0700 M NaCl on top. The transport numbers of K^+ and Na^+ in these solutions are 0.490 and 0.388, respectively. The boundary moved downward. The cross-sectional area of the tube was 0.100 cm2. The charge passed through was 96.5 coulombs.

(a) How far did the boundary move ?

(b) While the current was flowing, what was the concentration of NaCl just above the boundary ?

Solution : (a) Suppose the boundary moves a distance D for the passage of Q coulombs, then by definition,

$$D = \frac{\text{volume swept out by boundary}}{\text{cross-sectional area of the tube}}. \quad (1)$$

Before this relation can be used, the volume swept out by the boundary needs to be known.

The number of equivalents transported is $\frac{Q}{F}$, of which $\frac{t+Q}{F}$ are carried by the positive ion. Here t_+ is the fraction of the current carried by the positive ion in solution and is called the transport or transference number of the positive ion. C is the concentration in equivalents per cm^3, thus the volume swept out by the boundary during the passage of Q coulombs is calculated by the relation

$$V = \frac{t_K + Q/F}{C_K +} \quad (1)$$

In equation (2) $T_K +$ = transport number of K^+

C_K^+ = concentration of the K^+ ions in equivalents per cubic centimeter

As a result of fixing the units of CK^+, its value is 1.00×10^{-4} eq. cm^{-3}. This is obtained from 0.100 M KCl by the following :

$$0.100 \frac{\text{moles } K^+}{\ell} \times \frac{1 \text{ eq}}{\text{mole}} \times \frac{1\ell}{1000 \text{ ml}} \times \frac{1 \text{ ml}}{cm^3}.$$

F = Faraday's constant.

The Q/F term can be calculated as follows :

$$Q/F = \frac{96.5 \text{ coulombs}}{96500 \text{ coulombs Faraday}^{-1}}$$

$$= 1.00 \times 10^{-3} \cdot \text{Faraday} \times \frac{1 \text{ eq}}{\text{Faraday}} = 1\left(10^{-3}\right) \text{eq}.$$

Solving for V gives

$$V = \frac{(0.490)\left(1.00 \times 10^{-3}\right) \text{eq}}{\left(1.00 \times 10^{-4}\right) \frac{\text{eq}}{cm^3}}$$

$$= 4.90 \text{ cm}^3$$

If a = 0.100 cm^2 is the cross-sectional area of the tube, then the distance travelled by the boudary using equation (1) is

$$D = \frac{4.90 \text{ cm}^3}{0.100 \text{ cm}^2}$$

$$= 49.0 \text{ cm}$$

(b) While the current was flowing, the distance travelled by the K^+ ions is the same as the distance travelled by the Na^+ ions. Therefore, D is the same for both of them. Likewise, the volume swept out by the boundary remains the same value.

Thus equation (1) gives

$$4.90 = \frac{t_{NA} + Q/F}{C_{Na} +}$$

Note also that the Q/F remains the same. t_{Na+} is given to be 0.388. Therefore,

$$4.90 = \frac{(0.388)\left(1.00 \times 10^{-3}\right)}{C_{Na} +},$$

from which C_{Na^+} can be computed as follows :

$$C_{Na^+} = \frac{(0.388)\left(1.00 \times 10^{-3}\right)}{4.90}$$

$$= 7.92 \times 10^{-5} \text{ eq cm}^{-3}$$

or $$= 7.92 \times 10^{-2} \text{ mol liter}^{-1}.$$

Q. 2 : (a) A conductivity cell was calibrated using 0.01 N KCl ($k = 0.14087\ \Omega^{-1}\ m^{-1}$) in the cell and the measured resistance was 688 Ω. Find the cell constant.

(b) A 0.0100 N $AgNO_3$ solution in the same cell had a resistance of 777 Ω. What is Λ?

Solution : (a) The cell constant, L/S = kR (1)

where L = the distance between the electrodes

S = surface area of an electrode

k = specific conductance

R = resistance of the cell

$$L/S = (0.14087\ \Omega^{-1}\ m^{-1})\ (688\ \Omega) = 96.87\ m^{-1}$$

(b) $$\Lambda = \frac{k}{1000\ c} \qquad (2)$$

where c = concentration expressed in terms of equivalents dm^{-3} of solution (N, normality).

Λ = the equivalent conductance.

Solving for k in (1)

$$k = \frac{L/S}{R} = \frac{96.87\ m^{-1}}{777\ \Omega} = 0.1247\ \Omega^{-1}\ m^{-1}$$

Now using the equation Λ, in (2)

$$\Lambda = \frac{0.1247\ \Omega^{-1}\ m^{-1}}{\left(1000\ dm^{3}\ m^{-3}\right)\left(0.0100\ mol\ dm^{-2}\right)}$$

$$= 0.1247\ \Omega^{-1}\ m^{2}\ mol^{-1}$$

Q. 3 : Find the contributions of the conductivities of individual ions (ionic equivalent conductance λ_+ and λ_-) when Λ (equivalent conductance of electrolytic solution) = 0.0412 $m^2/(\Omega)$ (mol) and $t_+ = 0.825$.

Solution : By definition, $\Lambda = \lambda_+ + \lambda_-$ (1)

and $\Lambda t_i = \lambda_i$ (i = +, –) (2)

From (2) $\lambda^+ = [0.412\ m^2/(\Omega)\ (mol)]\ (0.825)$

$\lambda^+ = 0.03399\ m^2/(\Omega)\ (mol)$

Now using equation (1), $\lambda_- = \Lambda - \lambda_+$

$= 0.0412 - 0.03399$

$= 0.00721\ m^2/(\Omega)\ (mol)$

t_+ = transference (transport) number

Ω = resistance unit

Q. 4 : If 10.0 A were passed through a lead storage cell for 1.50 hr during a charging process, how much $PbSO_4$ would decompose ?

Solution : The balanced equation is

$$2PbSO_4(s) + 2H_2O(liq) = Pb(s) + PbO_2(s) + 2SO_4^{2-}(aq) + 4H^+(aq) \quad (1)$$

The number of coulombs of electricity passed through the cell is given by

$q = It$

where I = current (in amperes)

t = time (in seconds)

$$\therefore \quad q = (10\ A)\left(1.50\ hr \times \frac{3600\ secs.}{1\ hr}\right) = 5400\ coulombs$$

and this amount of electricity corresponds to

$$\frac{(5400\ coulombs)}{(96485\ coulombs/mol)} = 0.56\ mol$$

From the reaction equation of equation (1) the equivalent weight of $PbSO_4$ is equal to its molecular weight

= 303.25 gm/mol

= 0.30325 kg/mol

Recall that 0.56 mol correspond to the quantity of electricity.

the mass of $PbSO_4$ (S) reacting is then,

(0.56 mol) (0.30325 kg/mol) = 0.1697 kg.

Q. 5 : Given the two half-reactions, their standard reduction potentials (in volts) and the diagram below, what is the spontaneous reaction that occurs in the given cell and the voltage ?

Half-reaction	Standard reduction potential, E° (volts)
$Cu^{2+} + 2e^- \rightleftharpoons Cu$	0.34
$Zn^{2+} + 2e^- \rightleftharpoons Zn$	0.76

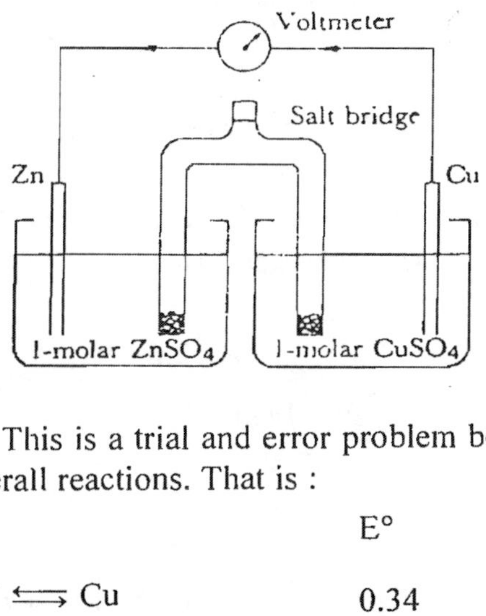

Solution : This is a trial and error problem because there are two possible overall reactions. That is :

Reactions	E°
$Cu^{2+} + 2e^- \rightleftharpoons Cu$	0.34
$Zn + Zn^{2+} + 2e^-$	0.76
$Cu^{2+} + Zn \rightleftharpoons Cu + Zn^{2+}$	1.10 volts

The reaction can proceed as written or in the reverse sense. Q = 1 since concentrations are equal. Therefore,

$$E = E° - \frac{RT}{nF} \ln 1 = E° = 1.10 \text{ volts.}$$

Since E° is positive, the reaction is spontaneous and must go in the forward direction (as written).

Q. 6 : The chemical reaction for the Daniell cell is

$$Zn\,(s) + Cu^{2+}\,(aq) = Zn^{2+}\,(aq) + Cu\,(s)$$

If $\Delta G_{25°C}$ (formation) = 0 for Zn (s) and Cu (S), –35.14 kcal mol^{-1} for Zn^{2+} (aq) and 15.66 kcal mol^{-1} for Cu^{2+} (aq), calculate the cell potential under standard conditions. Is the reaction spontaneous ?

Solution : $\Delta G°_{(reaction)} = -nFE°$. Therefore $E° = -\dfrac{\Delta G°_{(reaction)}}{nF}$

where $\Delta G°_{(reaction)}$ = standard free energy change of reactions

F = Faraday's Constant = 96.485 kJ $mol^{-1}\,V^{-1}$

n = number of moles of electrons in the balanced reactions

E° = cell potential

By definition, $\Delta G°_{25°C}$ (reaction) = $\Sigma n_i\, \Delta G°_{25°C}$ (products) – $\Sigma n_j\, \Delta G°_{25°C}$ (reactants)

where n_i and n_j = number of moles of products and reactants respectively.

$$\therefore\ \Delta G°_{25°C}\,(reaction) = [(1)(-35.14) + (1)(0)] - [(1)(0) + (1)(15.66)]$$

$$= -50.80\ kcal = (-50.80\ kcal)\,(4.186\ kJ/kcal)$$

$$= -212.6\ kJ.$$

Therefore
$$E° = \frac{-(-212.6\ kJ)}{(2\ mol)\left(96.485\ kJ\ mol^{-1}\ v^{-1}\right)}$$

$$= +\,1.102\ V$$

Since the cell potential is positive, the reaction is spontaneous as written.

Q. 7 : Find the cell potential for the cell whose reaction equation is $Ni + Sn^{2+} \rightarrow Ni^{2+} + Sn$ if Conc. Ni^{2+} = 1.3 and Conc. $Sn^{2+} = 1.0 \times 10^{-4}$. Also predict the direction of the spontaneous reaction.

Solution : The two half reaction and their standard potentials are :

	Reactions	E°
	$Ni \rightarrow Ni^{2+} + 2e^-$	+ 0.25
	$Sn^{2+} + 2e^- \rightarrow Sn$	– 0.14
Sum :	$Ni + Sn^{2+} \rightarrow Ni^{2+} + Sn$	E° = + 0.11

$$E = E° - \frac{0.059157}{n}\log Q \qquad (1)$$

where E = cell potential

$E°$ = standard potential for the two half reactions.

$$Q = \frac{(\text{concentration } Ni^{2+})}{(\text{concentration } Sn^{2+})} = \frac{1.3}{1.0 \times 10^{-4}} = 1.3 \times 10^4$$

n = number of electrons transferred = 2

Substituting values of E° and Q into equation (1), gives

$$E = 0.11 - \frac{0.059157}{2} \log\left(1.3 \times 10^4\right)$$

$$= 0.11 - (0.0296)(4.11) = 0.11 - 0.12$$

$$= -0.01 \text{ volts.}$$

The cell potential is negative, thus the reaction is not spontaneous as written. The criterion of spontaneous reaction is that E be positive.

Q. 8 : What is E° for the cell

$Pt / Ag\ (s) / AgCl\ (s) / Cl^-\ (a = 1) / Cl_2\ (1\ atm) / C\ (graph) / Pt$

if the standard half-cell reduction potentials are 0.2223 V for AgCl / Ag and 1.3583 V for Cl_2 / Cl^-? Is the reaction spontaneous?

Solution : Since E° for the cell is required, the individual E° for the half reactions are added together.

Both reactions are supposed to be reduction reactions. Therefore, the one with the smaller potential can be taken to be the oxidation reaction. In that case, the E° sign changes.

The two reactions are :

$$AgCl\ (s) + e^- \rightarrow Ag\ (s) + Cl^-\ (a = 1) \quad (1)$$

and

$$Cl_2\ (1\ atm) + 2e^- \rightarrow 2Cl^-\ (a = 1) \quad (2)$$

E° value for the first reaction is given to be 0.2223 V and that of the second reaction is 1.3583 V.

But it has been stated that the reaction with the smaller reduction potential will be written as the oxidation reaction. Therefore, reaction (1) is reversed and the sign of its potential changes. That is,

$$Ag\ (s) + Cl^-\ (a = 1) \rightarrow AgCl\ (s) + e^-$$

with an E° value of –0.2223 V.

Looking at both reactions, the first one contains an e^- and the second, a $2e^-$. Consequently, multiply the first by 2 and add to the second. The result is

$$2Ag\ (s) + Cl_2\ (1\ atm) \rightarrow 2AgCl\ (s).$$

Note that the potential is unaffected when multiplying reaction (1) by 2.

The cell potential is therefore (–0.2223) + (1.3583)

∴ E° for the cell = 1.1360 V.

A criterion for spontaneity (at standard conditions) is E° be positive. Therefore, the result indicates that the reaction is spontaneous.

Q. 9 : At 20°C the standard e.m.f. of the cell

$$Hg \mid Hg_2Cl_2\ (s),\ HCl\ (aq) \mid H_2,\ Pt$$

is 0.2692 V and at 30°C it is 0.2660 V. Find the values of $\Delta G°_m$, $\Delta H°_m$, and $\Delta S°_m$, at 25°C.

Solution : The two half reactions are

$$1/2\ Hg_2Cl_2\ (s) + e^- \rightarrow Hg\ (\ell) + Cl^-\ (aq)$$

and

$$1/2\ H_2\ (g) \rightarrow H^+\ (aq) + e^-$$

The overall cell reaction is obtained by adding up both equations. That is

$$1/2\ Hg_2Cl_2\ (s) + 1/2\ H_2\ (g) \rightarrow Hg\ (\ell) + Cl^-\ (aq) + H^+\ (aq).$$

By definition,

$$\Delta G°_m = -\ nFE°$$

where $\Delta G°_m$ = Standard Gibbs free energy

n = number of equivalents (which is one in this case because there is only one electron transfer)

F = Faraday's constant

E° = Standard e.m.f.

At 20°C = 293°C

$$\Delta G°_m = -\ (9.649 \times 10^4 C\ mol^{-1})\ (0.2692\ V)$$

$$= -25.98 \text{ kJ mol}^{-1}$$

and at 30°C = 303 °K

$$\Delta G^{\circ}_{m} = (9.649 \times 10^{4} \text{ C mol}^{-1})(0.2660 \text{ V})$$
$$= -25.67 \text{ kJ mol}^{-1}$$

Since 25°C is the average of 20°C and 30°C, ΔG°_{m} will also be the average ΔG°_{m} value at 20°C and 30°C. That is, at 25°C = 298 K

$$\Delta G^{\circ}_{m} = \frac{(-25.98) + (-25.67)}{2} = -\frac{51.65}{2}$$
$$= -25.825 \text{ kJ mol}^{-1}$$

from the above values, ΔS°_{m} can be approximated. By definition,

$$\Delta S^{\circ}_{m} = -\left(\frac{\partial \Delta G^{\circ}_{m}}{\partial T}\right)$$

But $$\left(\frac{\partial \Delta G^{\circ}_{m}}{\partial T}\right)_{298^{\circ}K} \sim \frac{-25.67 - (-25.98)}{(303 - 293)} = 31.00 \text{ JK}^{-1} \text{ mol}^{-1}$$

Therefore

$$\Delta S^{\circ}_{m} = -31.00 \text{ JK}^{-1} \text{ mol}^{-1}$$
$$\Delta H^{\circ}_{m} = \Delta G^{\circ}_{m} + T\Delta S^{\circ}_{m}$$

∴ At 298°K,

$$\Delta H^{\circ}_{m} = -25.82 + (298)(-31.00)$$
$$= -35.06 \text{ kJ mol}^{-1}$$

Q. 10 : What is the voltage at 25°C of the cell

Pt / H_2 (1 atm) / HCl (0.5 M) / HCl (1.0 M) / H_2 (1 atm) / Pt if $t_+ = 0.83$?

Solution : This is a concentration cell as specified by the molar quantities in parenthesis. Concentration cells depend for their emf on a transfer of material from one electrode to the other as a result of concentration difference between the two. This difference in concentration, as in the above cell, arises from the solutions with which the electrodes are in contact. The cell has a liquid junction between the two HCl solutions of the same kind but different concentrations, resulting in a concentration cell with transference, t.

In the above cell, the concentration, (1.0 M) of HCl in which the right electrode dips is greater than the concentration (0.5 M) of the same HCl in which the left electrode dips. It can thus be assumed that the left electrode is negative. The reaction at the negative electrode will be

$$1/2\ H_2\ (1\ \text{atm}) = H^+\ (0.5\ M) + e^-$$

and that at the positive electrode will be

$$H^+\ (1.0\ M) + e^- = 1/2\ H_2\ (1\ \text{atm})$$

Adding the two reactions together gives the total reaction

$$H^+\ (1.0\ M) = H^+\ (0.5\ M)$$

By constant, current generated by the Cl^- is moving from the right to left electrode and the positive ions, H^+, is moving across the junction from left to right.

The voltage (E) of this concentration cell is given by

$$E = -\frac{RT}{nF} \ln \left(\frac{(0.5\ M)^{t_-}}{(1.0\ M)^{t_-}} \right) \tag{1}$$

where t_- is the transference (transport) number of the chloride ions (that is the ions to which the electrodes are not reversible). This voltage results from the fact that the overall process of the cell is given by

$$t_-\ HCl\ (1.0\ M) = t_-\ HCl\ (0.5\ M)$$

Since $T_+ + t_- = 1$, thus $t_- = 1 - t_+ = 0.17$. From (1)

$$E = -t_- \frac{RT}{nF} \ln \left(\frac{(0.5\ M)}{(1.0\ M)} \right)$$

In this problem, n = 1 because only one mole is being transferred (see the net equation). Therefore,

$$E = -\frac{0.059157}{1} (0.17) \log \frac{0.5}{1.0}.$$

log (0.5/1.0) = – 0.30 and E = 0.0030 volts.

Q. 11 : For the cell H_2 (1 atm) | HCl | AgCl | Ag, E° = 0.2220 V at 298 K. If the measured E = 0.396 V, what is the pH of the HCl solution ? Cite any approximations made.

Solution : The two half reactions are

$1/2\ H_2 \rightarrow H^+ + e^-$ (oxidation)

$AgCl + e^- \rightarrow Ag + Cl^-$ (reduction)

OVERALL : $1/2\ H_2 + AgCl \rightarrow Ag + H^+ + Cl^{-1}$

Now, writing the Nernst equation for this overall reaction gives

$$E = E^\circ - \frac{RT}{nF} \ln\left(\frac{a_{H+}\, a_{Cl-}}{a_{H_2}^{1/2}}\right) \quad (1)$$

where E = cell e.m.f

E° = standard potential

R = Gas Constant

T = Temperature

F = Faraday's constant

a = Activity

n = 1 (number of equivalents of electrons in the balanced reaction)

At 25°C = 298°K,

$$\frac{RT}{nF} \ln \times = 0.5916 \log \times.$$

Therefore, equation (1) becomes

$$E = E^\circ - 0.05916 \log\left(\frac{a_{H+}\, a_{Cl-}}{a_{H_2}^{1/2}}\right)$$

$$= E^\circ - 0.05916 \left[\log a_{H+} + \log a_{Cl-} - 1/2 \log a_{H_2}\right] \quad (2)$$

a_{H_2} can be approximated by P_{H_2} = 1 atm where P = pressure.

Therefore a_{H_2} = 1 and equation (2) becomes

$$E = E^\circ - 0.05916\ [\log a_{H+} + \log a_{Cl-}] \quad (3)$$

Since $a_{H+} = a_{Cl-}$, equation (3) can be written in the form

$$E = E° - (0.05916)(2) \log a_{H+}. \quad (4)$$

By definition, $PH = -\log a_{H+}$

Therefore, equation (4) becomes

$$E = E° + (0.05916)(2) \text{ pH} \quad (5)$$

Both values of E and E° are given to be 0.396 V and 0.2220 V respectively. Substitute these values into equation (5) to get

$$0.396 \text{ V} = 0.2220 \text{ V} + (0.05916)(2) \text{ pH}$$

or
$$0.396 - 0.2220 = +(0.05916)(2) \text{ pH}$$

$$0.174 = 0.11832 \text{ pH}$$

$$\therefore \quad \text{pH} = \frac{0.174}{0.11832} = 1.47$$

$$= 1.47$$

Q. 12 : The measured voltage of the concentration cell Ag | $AgNO_3$ (0.0100 m) || $AgNO_3$ (y) | Ag was 0.0650 V at 25°C. Calculate the molality y of the right-hand half-cell assuming $\gamma_\pm = 1$.

Solution : (a) The cell's overall reaction is given by

$$Ag^+ (y) \rightarrow Ag^+ (0.0100 \text{ m}).$$

For this cell, the Nernst equation is

$$E = E° - \frac{RT}{nF} \ln\left(\frac{a_\pm (0.01)}{a_\pm (y)}\right) \quad (1)$$

Here, $n = 1$

a = activity

E° = standard potential of the cell

F = Faraday's constant

Because of the similarity of the reactions in a concentration cell, E° = 0 and the potential is given by the non-standard conditions of the reactants and products. Therefore, equation (1) becomes

$$E = -\frac{RT}{F} \ln\left(\frac{a_\pm (0.01)}{a_\pm (y)}\right) \quad (2)$$

By definition, the mean activity of a solute undergoing ionization in solution is given by

$$a_{\pm} = \gamma_{\pm} C_{\pm}$$

where $\gamma_{\pm}$ = mean activity coefficient

$C_{\pm}$ = mean ionic molal concentration

But $\gamma_{\pm}$ is given to be 1, therefore, $a_{\pm} + C_{\pm}$ and equation (2) becomes

$$E = -\frac{R(298)}{F} \ln\left(\frac{0.01}{y}\right) = -0.05916 \log\left(\frac{0.01}{y}\right) \text{ at } 25°C.$$

or $$\log\left(\frac{0.01}{y}\right) = -\frac{E}{0.05916} = -\frac{0.0650}{0.05916} = -1.0987;$$

From which $$\frac{0.01}{y} = 7.967 \times 10^{-2}$$

and $$y = \frac{0.01}{7.967 \times 10^{-2}}$$
$$= 0.1255 \text{ m}$$

Q. 13 : For the reaction $PbSO_4$ (s) $\rightarrow Pb^{2+} + SO_4^{2-}$, $\Delta H = +2990$ cal/mol. Will the solubility of $PbSO_4$ increase or decrease with increasing temperature? $K_{sp} = 1.8 \times 10^{-8}$ at 25°C. Find its k_{sp} at 55°C.

Solution: Whether the solubility of a salt increases or decreases with an increase in temperature can be determined by an investigation of its ΔH, the enthalpy change. The formula for the temperature dependence on k is

$$\log k = \frac{\Delta H°}{2.303\ RT} + \frac{\Delta S°}{2.303\ R}.$$

A positive $\Delta H°$ suggests that solubility increases with increasing temperature. When $\Delta H°$ is positive, the first term remains a negative value. When T increases, the first term becomes a smaller negative number. Thus, k increases. Since k measures solubility, the solubility increases. To find the k_{sp} of $PbSO_4$ at 55°C, use the fact that

$$\log \frac{k_2}{k_1} = \frac{-\Delta H°}{2.303\ R}\left[\frac{1}{T_2} - \frac{1}{T_1}\right],$$

where k_2 = solubility constant at the temperature in Kelvin (Celsius plus 273°) of T_2; k_1 = solubility constant at a temperature of T_1. Thus, if the k_{sp} at one temperature is known, the k_{sp} at another temperature can be found given ΔH°. Let T_2 = 55° C or 328°K and T_1 = 298°K. (25°C). Thus

$$\log \frac{k_{328}}{k_{298}} = \frac{-\Delta H°}{2.303\ R}\left[\frac{1}{328} - \frac{1}{298}\right]$$

$$\log \frac{k_{328}}{1.8 \times 10^{-8}} = \frac{-(2990\ \text{cal / mole})(-3.07 \times 10^{-4}\ /\ \text{K})}{(2.303)\ 1.987\ \text{cal / mole K}}$$

$$= 0.200$$

$$k_{328} = (1.8 \times 10^{-8})\ (10^{0.200})$$

$$= (1.8 \times 10^{-8})\ (1.58)$$

$$= 2.8 \times 10^{-8}.$$

Q. 14 : Calculate ΔE° for the following cells : (1) Cadmium and Hydrogen, (2) Silver and Hydrogen and (3) Cadmium and Silver, using the following data :

Reaction	E° volts
$Cd \rightarrow Cd^{+2} + 2e^-$	+ .403
$H_2 \rightarrow 2H^+ + 2e^-$	0.00
$Ag \rightarrow Ag^+ + e^-$	− .799

Solution : You are asked to calculate the standard cell potential (ΔE°) for each of the given pairs. To do this, you must realize that in such cells, you have 2 half-reactions. Namely, an oxidation reaction (loss of electrons) and a reduction reaction (gain of electrons). The sum of these half-reactions yields the overall reaction and the ΔE° of the whole cell. Thus, to find the ΔE° for each of these pairs, you need to know the E° of the half-reactions. There is one other important fact to be kept in mind.

In cells, if the reaction is to proceed spontaneously, $\Delta E°$ must have a positive value. This means, therefore, that you must choose the half-reactions such that their sum always gives a positive $\Delta E°$. You are told the E° of the oxidation half-reaction for each element in each pair. To reduction half-reaction is the reverse of the oxidation reaction for each element, with a change in sign of E°. For example, if you have $A \rightarrow A^{+1} + e^-$ with an E° = B for oxidation, the reduction is $A^{+1} + e^- \rightarrow A$ with a E° = – B.

With this in mind, the procedure is as follows :

(1) Cadmium (Cd) and hydrogen (H_2).

The reaction for this cell must be the sum of the oxidation and reduction such that the $\Delta E°$ is positive. This can only occur if the anode (oxidation) has the higher oxidation potential. Thus, you calculate $\Delta E°$ as + .403 – (0.000) = + .403 V. Similarly, for (2) and (3), a positive $\Delta E°$ can only be obtained with the anode having the higher oxidation potential. Thus, for (2), E° = (0.000) – (– .799) = + .799v. For (3), E° = (.403) – (– .799) = + 1.202v.

Q. 15 : For the following voltaic cell, write the half-reactions, designating which is oxidation and which is reduction. Write the cell reaction and calculate the voltage of the cell made from standard electrodes. The cell is Co; Co^{+2} || Ni^{+2}; Ni.

Solution : The cell reaction is the algebraic sum of the reactions that take place at the electrodes. Every cell has 2 electrodes an anode and a cathode. Oxidation, which is the loss of electrons, occurs at the anode. Reduction, which is the gain of electrons, take place at the cathode.

The cell is always written as solid; ion in solution || ion in solution; solid (anode) (cathode)

Oxidation and reduction are the half reactions that take place in the cell. For this cell, they are

$$Co \rightarrow Co^{+2} + 2e^- \quad \text{(oxidation at anode)}$$

$$Ni^{+2} + 2e^- \rightarrow Ni \quad \text{(reduction at cathode).}$$

$$\text{Sum : } Ni^{+2} + Co \rightarrow Ni + Co^{+2} \quad \text{(Cell reaction).}$$

Since Co is losing electrons, it provides the oxidation reaction and Ni^{+2}, gaining these electrons, takes part in the reduction reaction.

The voltage of a cell is the sum of the oxidation and reduction potentials in units of volts and is designated by E° (under standard conditions).

$$E^\circ_{cell} = E^\circ_{oxidation} + E^\circ_{reduction}.$$

The voltages of half cell reactions are usually given as the reduction potentials. The oxidation potential is opposite in sign to the reduction potential. The potentials can be obtained from a table of standard reduction potentials.

$$Ni^{+2} + 2e^- \rightarrow Ni, \text{ the potential is } -.25 \text{ v, } E^\circ_{red} = -.25 \text{ v.}$$

For $Co^{+2} + 2e^- \rightarrow Co$, the potential is –.277 v.

Since $Co \rightarrow Co^{+2} + 2e^-$ is the oxidation reaction, E°_{ox} equals the negative of –.277 v, or $E^\circ_{ox} = .277$ v.

Substituting these values into the equation $E^\circ = E^\circ_{oxid} + E^\circ_{red}$, one obtains

$$E^\circ = +.277 + (-.25) = .027 \text{ v.}$$

Since E° is positive, the reaction proceeds spontaneously and can be used to supply current.

Q. 16 : For the following voltaic cell, write the half reactions, designating which is oxidation and which reduction. Write the cell reaction and calculate the voltage (E°) of the cell from the given electrodes. The cell is

$$Cu; Cu^{+2} \,||\, Ag^{+2}; Ag.$$

Solution : In a voltaic cell, the flow of electrons creates a current. Their flow is regulated by 2 types of reactions occur concurrently, oxidation and reduction. Oxidation is a process where electrons are lost and reduction where electrons are gained. The equation for these are the half-reactions. From the cell diagram, the direction of the reaction is always left to right.

$$Cu \rightarrow Cu^{+2} + 2e^- \qquad \text{oxidation}$$

$$Ag^{+2} + 2e^- \rightarrow Ag \qquad \text{reduction.}$$

Therefore, the combined cell reaction is

$$Cu + Ag^{+2} \rightarrow Cu^{+2} + Ag.$$

To calculate the total E°, look up the value for the E° of both half-reactions as reductions. To obtain E° for oxidation, reverse the sign of the reduction E°. Then, substitute into $E°_{cell} = E°_{red} + E°_{ox}$. If you do this, you find

$$E°_{cell} = -(E°_{red}\ Cu) + E°_{red}\ Ag^{+2}$$
$$= -.34 + .80$$
$$= .46 \text{ volt.}$$

Q. 17 : Given the following standard electrode potentials at 25°C: $Sn^{4+} + 2e^- \rightarrow Sn^{2+}$, E° = 0.15 ev and $Fe^{3+} + e^- \rightarrow Fe^{2+}$, E° = + 0.77ev wi!l the reaction $Sn^{2+} + 2Fe^{3+} \rightleftarrows Sn^{4+} + 2Fe^{2+}$ proceed spontaneously ?

Solution : A reaction will proceed spontaneously only if it has a positive ΔE° as written. Thus, to answer this question, you want to calculate the ΔE° and see whether it is positive. This can be done by considering its half-reactions. The sum of the electrode potentials for these reactions will be the ΔE°.

The overall reaction is

$$Sn^{2+} + 2Fe^{3+} \rightleftarrows Sn^{4+} + 2Fe^{2+}.$$

The half-reactions are, oxidation (loss of electrons) and reduction (gain of electrons) reactions. You have then,

$$\text{oxidation : } Sn^{2+} \rightarrow Sn^{4+} + 2e^-.$$

This, has the reverse of the standard potential of the reaction given. Thus, its E° = – (0.15) = – .15e v.

$$\text{Reduction : } Fe^{3+} + e^- \rightarrow Fe^{2+}.$$

This has the same standard potential as in the given reaction. Thus, its E° = 0.77ev. Adding these two equations together, one obtains the desired net reaction.

$$S_n^{2+} \rightarrow Sn^{4+} + 2e^-$$
$$2Fe^{3+} + 2e^- \rightarrow 2Fe^{2+}$$

$$2Fe^{3+} + Sn^{2+} \rightarrow Sn^{4+} + 2Fe^{2+}$$

The $\Delta E° = E°_{oxid} + E°_{red} = -.15 + .77 = +.62$ volt. The fact that it is positive means the reaction proceeds spontaneously.

Q. 18 : Calculate the voltage of the cell Fe; Fe^{+2} || H^+; H_2 if the iron half cell is at standard conditions but the H^+ ion concentrations is .001M.

Solution : The voltage (E) of a cell is found using the Nernst Equation because it involves the use of concentration factors. It is stated

$$E = E° - \frac{RT}{NF} \ln Q,$$

where R is 8.314 joules per degree, F is 96,500 coulombs, N is the number of moles of electrons transferred and Q is the concentration term. T = 25° C, by definition of standard condition, in this equation (or 298° K).

But to solve the problem we must first obtain E°. This is done by writing down the appropriate half-reactions. Oxidation is the loss of electrons and reduction is the gain of electrons.

Reaction	Type	$E°_{red}$
$Fe \rightarrow Fe^{+2} + 2e^-$	Oxidation	+ .44
$2H^+ + 2e^- \rightarrow H_2$	Reduction	0

Next, take the algebraic sum of the $E°_{red}$ and $E°_{oxid}$, which gives E°.

$$E° = + .44 + 0 = + .44.$$

Now set up the concentration term.

$$\ln Q = \ln \frac{[H^+]^2}{[Fe^{+2}]} = \ln \frac{[.001]^2}{[1]}$$

Standard conditions always means a concentration of 1 M. Substituting all these terms into the Nernst equation, one claculates E to be

$$E = + .44 - \frac{.059}{2} \ln \frac{10^{-6}}{1} = .85 \text{ volt.}$$

Q. 19 : Calculate the voltage (E) of a cell with E° = 1.1 volts, if the copper half-cell is at standard conditions but the zinc ion concentration is only .001 molar. Temperature is 25°C. The overall reaction is

$$Zn + Cu^{+2} \rightarrow Cu + Zn^{+2}.$$

Solution : To solve this problem requires a relation between the cell potential (voltage), E, and the standard cell potential, E°. The equation used is the Nernst equation, which states

$$E = E° - \frac{RT}{nF} \ln (Q),$$

where n is the number of mole of electrons transferred, R is the constant 8.314 joules per degree, F is 96,500 coulombs, and Q is the equilibrium constant expression. In other words,

$$Q = \frac{[\text{product}]}{[\text{reactant}]}$$

You are told the concentration of Zn^{+2} is .001.

$$= \frac{[Zn^{+2}]}{[Cu^{+2}]} = \frac{.001}{1}$$

Standard conditions indicate that the concentration of the substance is 1M and at a temperature of 25°C. Thus $[Cu^{+2}] = 1$ and therefore, from the reaction equation, it can be seen that 2 electrons are transferred. Thus, n = 2.

Substituting these values into the Nernst equation, one obtains

$$E = 1.1 - \frac{8.314\ (298)}{2\ (96{,}500)} \log \frac{.001}{1}$$

$$= 1.1\ - .0295\ (-3)$$

$$= 1.19 \text{ volts.}$$

Q. 20 : Calculate the solubility product constant of pure $PbSO_4$ in water. The solubility of $PbSO_4$ in water at 25°C is 1.25×10^{-4} moles/ liter.

Solution : The solubility of a substance is the number of moles of that substance which will dissolve in one liter. Since $PbSO_4$ contains one Pb^{2+} ion and one $SO_4^{\ 2-}$ ion per molecule in aqueous solution, the concentration of each of these ions is equal to the solubility. Hence,

$$[Pb^{2+}] = [SO_4^{2-}] = 1.25 \times 10^{-4} \text{ M.}$$

The solubilization reaction for $PbSO_4$ is

$$PbSO_4\,(s) \rightleftarrows Pb^{2+} + SO_4^{2-}$$

and the solubility product constant is

$$K_{sp} = \left[Pb^{2+}\right]\left[SO_4^{2-}\right].$$

Substituting the values for $\left[Pb^{2+}\right]$ and $\left[SO_4^{2-}\right]$ gives

$$K_{sp} = (1.25 \times 10^{-4}\ M) \times (1.25 \times 10^{-4}\ M)$$
$$= 1.6 \times 10^{-8}\ \text{moles}^2/\text{liter}^2.$$

Q. 21 : Calculate the minimum concentration of Br^- ion necessary to bring abou⁺ precipitation of AgBr from a solution in which the concentration Ag^+ ion is 1×10^{-5} mole per liter.

$$K_{sp} \text{ for AgBr} = 4 \times 10^{-13}.$$

Solution : The product of the concentrations of ions in a saturated solution of a relatively insoluble salt, such as AgBr, at a given temperature is constant. This constant is called the solubility product constant (K_{sp}). For AgBr, K_{sp} can be stated as

$$K_{sp} = [Ag^+][Br^-] = 4 \times 10^{-13}.$$

In this problem, we are given the concentration of Ag^+ as 1×10^{-5} M, therefore $[Br^-]$ can be found by using the equation for K_{sp}.

$$\left[Ag^+\right]\left[Br^-\right] = 4 \times 10^{-13}$$

$$\left[1 \times 10^{-5}\right]\left[Br^-\right] = 4 \times 10^{-13}$$

$$\left[Br^-\right] = \frac{4 \times 10^{-13}}{1 \times 10^{-5}} = 4 \times 10^{-8}\ M.$$

If the concentration of Br^- ions is raised to 4×10^{-8} M, AgBr will precipitate out, for the addition of any more Br^- ions could not be supported by the solubility of the ions in solution, as the K_{sp} indicates.

Q. 22 : Will precipitation occur if .01 mole of Ba^{+2} is added to a liter of solution containing .05 mole of SO_4^{-2}? K_{sp} for $BaSO_4 = 1 \times 10^{-10}$.

Solution : K_{sp} is the solubility production constant. It is defined as the minimum product of the concentration of the ions needed to bring about the formation of a precipitate (i.e., a solid compound). The general form of the K_{sp} equation for compound $A_n B_m$ is :

$$K_{sp} = [A]^n [B]^m$$

Therefore, for $BaSO_4$ the K_{sp} equation is :

$$K_{sp} = \left[Ba^{+2}\right]\left[SO_4^{-2}\right] = 1 \times 10^{-10}.$$

To find out if precipitation occurs, substitute in the given concentration values. If the calculated value is greater that the given value for K_{sp}, then precipitation occurs. If it is lower, then no precipitation occurs. Thus,

$$K_{sp} = \left[Ba^{+2}\right]\left[SO_4^{-2}\right]$$
$$= \left[1 \times 10^{-2}\right]\left[5 \times 10^{-2}\right]$$
$$= 5 \times 10^{-4}$$

Since this value is greater than the given K_{sp} value, precipitation occurs.

Q. 23 : Find K_b and pK_b for the acetate ion CH_3COO^-. The ionization constant of CH_3COOH is $K_a = 1.75 \times 10^{-5}$; $K_W = 1.00 \times 10^{-14}$.

Solution : This problem is concerned with conjugate acid-base pairs. In the general reaction,

$$\underset{\text{acid}}{HA} + \underset{\text{base}}{H_2O} \rightleftarrows \underset{\text{acid}}{H_3O^+} + \underset{\text{base}}{A^-}$$

HA is the conjugate acid of A–, or conversely, A^- is the conjugate base of HA. H_2O has acted as a week base. The acid and base ionization constants for a conjugate pair, all illustrated above, are K_a and K_b.

$$K_a K_b = K_W \text{ which means } pK_a + pK_b = pK_W.$$

Thus, if given the ionization constant for an acid or base, one can calculate the ionization constant for its conjugate base or acid.

Also, remember that the K and pK values are related by the general equation,

$$pK = -\log K.$$

Therefore, the values of K_b and pK_b are

$$K_b = \frac{K_W}{K_a} = \frac{1.00\times10^{-14}}{1.75\times10^{-5}} = 5.71\times10^{-10}.$$

$$pK_b = -\log K_b = -\log 5.71\times10^{-10}$$
$$= -[-10 + \log 5.71] = 9.244.$$

Q. 24 : Find the pH of a 0.2 M solution of formic acid. $K_a = 1.76 \times 10^{-4}$.

Solution : To solve the problem, first note that the acid dissociates according to the general equation

$$HA + H_2O \rightleftarrows H^+ + A^-$$

where HA is the acid and H^+ and A^- are the dissociation products. To find the pH, calculate $[H^+]$.

One can see from the equation that equal concentrations of H^+ and A^- form when HA dissociates. At equilibrium, x moles of HA dissociates into x moles of H^+ and x moles of A^-.

	HA + H_2O ⇄	H^+ +	A^-
Before :	0.2	0	0
After :	0.2 – x	x	x

Now, set up the equilibrium constant equation, which is

$$K_a = \frac{[\text{products}]}{[\text{reactants}]} = \frac{[H^+][A^-]}{[HA]} = 1.76 \times 10^{-4}.$$

Before substituting in the values, simplify the calculation by assuming that $0.2 - x \cong 0.2$. This is a valid assumption since the acid is very weak and only a very small percentage dissociates. Thus, one has

$$K_a = \frac{(x)(x)}{(0.2)} = 1.76 \times 10^{-4}. \quad \text{Solving :}$$

$$x = [H^+] = 5.93\times10^{-3}.$$

H^+ concentration is then converted to pH according to the equation,

$$pH = -\log[H^+] = -\log 5.93 \times 10^{-3}$$
$$= -[-3 + \log 5.93] = 3 - 0.77 = 2.23.$$

Q. 25 : Find the pH of a 0.1 M solution of ammonia, NH_3. $pK_b = 1.76 \times 10^{-5}$.

Solution : This problem involves the dissociation of a weak base, which proceeds by the following general equation :

$$B + H_2O \rightleftarrows BH^+ + OH^-,$$

where B is the base and BH^+ and OH^- are the dissociation products. x moles of B dissociate into x moles of BH^+ and x moles of OH^-. This is more clearly seen below :

	$NH_3 + H_2O \rightleftarrows$	NH_4^+	$+ OH^-$
Before :	0.1 M	O	O
After :	0.1 – x	x	x

If x moles of NH_3 dissociate, then at equilibrium, one has 0.1 – x left, since 0.1 is the initial concentration. NH_3 is a weak base and very little dissociation occurs, which means x is extremely small. This allows one to approximate 0.1 – x as 0.1. By now, it should be evident that $[H^+]$ does not appear in the equilibrium equation. The strategy will be to first find $[OH^-]$, then calculating the pOH, and then using the equation pK = pOH + pH = 14 to find pH.

In order to calculate the value of $[OH^-]$, set up the equilibrium constant equation for which $K_b = 1.76 \times 10^{-5}$.

Thus, $$K_b = \frac{[NH_4^+][OH^-]}{[NH_3]}.$$ Substituting,

$$K_b = \frac{(x)(x)}{0.1} = 1.76 \times 10^{-5}, \text{ so that}$$

$$x = [OH^-] = 1.33 \times 10^{-3}.$$

Then, one can calculate pOH by :

$pOH = -\log [OH^-] = -\log [1.33 \times 10^{-3}]$

$= 3 - \log 1.33 = 2.88.$

Therefore, $pH = 14 - pOH = 14 - 2.88 = 11.12.$

Q. 26 : (a) Determine the pH of a solution with a hydrogen ion concentration of 3.5×10^{-4}.

(b) If a solution has a pH of 4.25, what is the hydrogen ion concentration ?

Solution : To determine the acidity or basicity of an aqueous solution, the hydrogen ion concentration must be measured. The pH of a solution express this concentration. pH is defined as the negative logarithm of the hydrogen ion concentration. In other words, pH = $(-\log [H^+])$, where the brackets around H^+ signify concentration. As such, to solve the problem, you must substitute into the equation. For part "A", you have $pH = -\log [3.5 \times 10^{-4}]$.

now $-\log [3.5 \times 10^{-4}] = -\log 3.5 - \log 10{-}4$

$= -.54 - (-4)$

$= -.54 + 4$

$= 3.46$

If follows, then, that the pH = 3.46 for a hydrogen ion concentration of 3.5×10^{-4}. Part "B" is similar, but here you are given the pH and asked to find the ion concentration. Therefore, you have $4.25 = -\log [H^+]$ or $-4.25 = \log [H^+]$. Now, logarithm numbers give only positive mantissas. As such, – 4.25 must be in the form of –5 + .75. If you take the antilogarithm of each, .75 is 5.6 and –5 is 10^{-5}, you obtain a hydrogen ion concentration of 5.6×10^{-5} mole/liter.

Q. 27 : The ionization constant for NH_4OH is 1.8×10^{-5}.

Calculate the concentration of OH^- ions in a 1.0 molar solution of NH_4OH.

Solution : The ionization constant (K_b) is defined as the concentration of OH^- ions times the concentration of the conjugate acid ions of a given base devided by the concentration of unionized base. For a base, BA,

$$K_b = \frac{[B^-][A^+]}{[BA]},$$

where K_b is the ionization constant, $[B^-]$ is the concentration of ionized base ions, $[A^+]$ is the concentration of the conjugate acid, and [BA] is the concentration of unionized base. The K_b for NH_4OH is stated as

$$K_b = \frac{[NH_4^+][OH^+]}{NH_4OH} = 1.8 \times 10^{-5}$$

When NH_4OH is ionized, one NH_4^+ ion is formed and one OH^{-1} ion is formed,

$$NH_4OH \rightleftarrows NH_4^+ + OH^-$$

Thus, the concentrations of each ion are equal.

$$[NH_4^+] = [OH^-]$$

The concentration of unionized base is decreased when ionization occurs. The new concentration is equal to the concentration of OH^- subtracted from the concentration of NH_4OH.

$$[NH_4OH] = 1.0 - [OH^-]$$

Since $[OH^-]$ is small relative to 1.0, one may assume that $1.0 - [OH^-]$ is approximately equal to 1.0

$$[NH_4OH] = 1.0 - [OH^-] \cong 1.0$$

Using this assumption, and the fact that $[OH^-] = [NH_4^+]$, K_b can be rewritten as

$$K_b = \frac{[OH^-][OH^-]}{1.0} = 1.8 \times 10^{-5}$$

Solving for $[OH^-]$:

$$\frac{[OH^-][OH^-]}{1.0} = 1.8 \times 10^{-5}$$

$$[OH^-]^2 = 1.8 \times 10^{-5}$$

$$[OH^-]^2 = \sqrt{1.8 \times 10^{-5}} = 4.2 \times 10^{-3}.$$

14
Nuclear Chemistry

Q. 1 : The radioactive decay constant for radium is 1.36×10^{-11}. How many disintegrations per second occur in 100 g of radium ?

Solution : The number of disintegrations per second of a given amount of a particular element can be determined by using the following equation.

$$D = \lambda N$$

where D is the number of disintegrations per second, λ the decay constant and N the number of atoms present.

In this problem one is given λ and must determine N before solving for D. One is told that 100 g of radium is present, the number of moles present is determined by dividing 100 g by the molecular weight of radium, 226 g/mole.

$$\text{no. of moles} = \frac{100 \text{ g}}{226 \text{ g / mole}} = 0.442 \text{ moles}$$

There are 6.02×10^{23} particles per mole, thus, one can calculate the number of atoms (N) in 0.442 moles.

$$N = (0.442 \text{ moles}) (6.02 \times 10^{23} \text{ atoms/mole})$$
$$= 2.66 \times 10^{23} \text{ atoms}$$

Solving for D :

$$D = \lambda N = (1.36 \times 10^{-11}) (2.66 \times 10^{23}) = 3.62 \times 10^{12} \text{ dis/sec.}$$

Q. 2 : Calculate ΔE for the proposed basis of an absolutely clean source of nuclear energy

$$^{11}_{5}B + ^{1}_{1}H \rightarrow ^{12}_{6}C \rightarrow 3\ ^{4}_{2}He$$

Atomic masses : $^{11}B = 11.00931$, $^{4}He = 4.00260$, $^{1}H = 1.00783$.

Solution : The energy of nuclear reactions, ΔE, is calculated from the difference between the masses of products and reactants in accordance with the Einstein Law. Einstein's Law can be stated $\Delta E = \Delta mc^2$, where Δm is the difference in the masses of the products and the reactants, and c is the speed of light (3×10^{10} cm/sec). The total reaction here can be written

$$^{11}_{5}B + ^{1}_{1}H \rightarrow 3\ ^{4}_{2}He.$$

Thus, Δm is equal to the mass of $^{11}_{5}B$ and $^{1}_{1}H$ substracted from the mass of $3\ ^{4}_{2}He$. The mass of $3\ ^{4}_{2}He$ is equal to 3 times the mass of $^{4}_{2}He$.

$$\begin{aligned}\Delta m &= \left(3 \times \text{m of } ^{4}_{2}He\right) - \left(\text{m of } ^{11}_{5}B + \text{m of } ^{1}_{1}H\right)\\ &= (3 \times 4.00260) - (11.00931 + 1.00783)\\ &= 12.0078 - 12.01714 = -9.34 \times 10^{-3}\ \text{g/mole.}\end{aligned}$$

One can now solve for ΔE by using Δm.

$$\begin{aligned}\Delta m = \Delta mc^2 &= -9.34 \times 10^{-3}\ \text{g/mole} \times \left(3 \times 10^{10}\ \text{cm/sec}\right)^2\\ &= -9.34 \times 10^{-3}\ \text{g/mole} \times 9.0 \times 10^{20}\ \text{cm}^2/\text{sec}^2\\ &= -8.406 \times 10^{18}\ \text{g cm}^2/\text{mole sec}^2\\ &= -8.406 \times 10^{18}\ \text{ergs/mole.}\end{aligned}$$

There are 4.18×10^{18} ergs in 1 Kcal, thus ergs can be converted to Kcal by dividing the number of ergs by the conversion factor 4.18×10^{10} ergs / Kcal.

$$\begin{aligned}\text{no. of Kcal} &= \frac{-8.406 \times 10^{18}\ \text{ergs/mole}}{4.18 \times 10^{10}\ \text{ergs/Kcal}}\\ &= -2.01 \times 10^{8}\ \text{Kcal/mole.}\end{aligned}$$

ΔE for this reaction is -2.01×10^{8} Kcal per mole.

Q. 3 : A chemist is given an unknown element X. He finds that the element X has an atomic weight of 210.197 amu and consists of only two isotopes, ^{210}X and ^{212}X. If the masses of these isotopes are, respectively, 209.64 and 211.66 amu, what is the relative abundance of the two isotopes ?

Solution : The relative abundance of the two isotopes is equal to their fraction in the element. The sum of the fractions of the isotopes times their respective masses is equal to the total atomic weight of X. The element X is composed of only two isotopes. The sum of their fractions must be equal to one.

Solving : Let y = the fraction of the 210 isotope. Since, the sum of the fractions is one, 1 – y = the fraction of the 212 isotope. The sum of the fractions times their masses, equals the atomic weight of X.

$$209.64\, y + 211.66\,(1 - y) = 210.197;$$

$y = 0.742$ = fraction of ^{210}X. $1 - y = 0.2758$ = fraction of ^{212}X.

By percentage the relative abundance of ^{210}X is 72.42% and ^{212}X is 27.58%.

Q. 4 : Natural chlorine is a mixture of isotopes. Determine its atomic weight if 75.53% of the naturally occurring element is chlorine 35, which has a mass of 34.968, and 24.47% is chlorine 37, which has a mass of 36.956.

Solution : If naturally occurring chlorine contains 75.53% chlorine 35 and 24.47% chlorine 37, one mole of chlorine contains .7553 mole chlorine 35 and .2447 mole chlorine 37. The weight of one mole of chlorine is, then, equal to the sum of the weight of .7553 mole ^{35}Cl and .2447 mole ^{37}Cl.

$$\text{atomic weight of Cl} = (.7553)(34.968) + (.2447)(36.956)$$

$$= 26.34 + 9.11 = 35.45.$$

Q. 5 : A chemist is given an unknown element X. He finds that the element X has an atomic weight of 210.197 amu and consists of only two isotopes, ^{210}X and ^{212}X. If the masses of these isotopes are, respectively, 209.64 and 211.66 amu, what is the relative abundance of the two isotopes ?

Solution : The relative abundance of the two isotopes is equal to their fraction in the element. The sum of the fractions of the isotopes times their respective masses is equal to the total atomic weight of X. The element X is composed of only two isotopes. The sum of their fractions must be equal to one.

Solving : Let y = the fraction of the 210 isotope. Since the sum of the fractions is one, 1 – y = the fraction of the 212 isotope. The sum of the fractions times their masses, equals the atomic weight of X.

$$209.64\,y + 211.66\,(1 - y) = 210.197;$$

$$209.64\,y + 211.66 - 211.66\,y = 210.197$$

$$-2.02\,y = -1.463$$

$$y = 1.463/2.02$$

$$y = 0.7242 = \text{fraction of } {}^{210}X.$$

$$\therefore \quad 1 - y = 0.2758 = \text{fraction of } {}^{212}X.$$

By percentage the relative abundance of ${}^{210}X$ is 72.42% and ${}^{212}X$ is 27.58%.

Q. 6 : Chromium exists in four isotopic forms. The atomic masses and percent occurrences of these isotopes are listed in the following table :

Isotopic mass (amu)	Percent occurrence
50	4.31%
52	83.76%
53	9.55%
54	2.38%

Calculate the average atomic mass of chromium.

Solution : By definition,

$$A = \sum_{i=1}^{N} p_i M_i, \qquad (1)$$

where A = average value, pi = probability of occurence of isotope i

Mi = atomic mass of isotope i. Making use of the above definition and the given data, the average atomic mass of chromium can be computed.

For the four isotopes of chromium,

$$M_1 = 50 \text{ amu } p_1 = 4.31\% = 0.0431$$

$$M_2 = 52 \text{ amu } p_2 = 83.76\% = 0.8376$$

$$M_3 = 53 \text{ amu } p_3 = 9.55\% = 0.0955$$

$$M_4 = 54 \text{ amu } p_4 = 2.38\% = 0.0238$$

Hecen, the average atomic mass of chromium is $A = p_1M_1 + p_2M_2 + p_3M_3 + p_4M_4 = 0.0431 \times 50$ amu $+ 0.8376 \times 52$ amu $+ 0.0955 \times 53$ amu $+ 0.0238 \times 54$ amu $= 2.155$ amu $+ 43.555$ amu $+ 5.062$ amu $+ 1.285$ amu $= 52.057$ amu.

Q. 7 : Calculate the total and average binding energies and the packing fraction for ${}^{12}_{6}C$.

Solution : The general form of representing a particular nuclear reaction is given by

$$ {}^{A}_{Z}X = Z\,{}^{1}_{1}H + (A - Z)\,{}^{1}_{0}n \qquad (1)$$

where X = elemental chemical symbol corresponding to the value of Z.

Z = atomic number

A = protons + neutrons = mass number.

Consequently, for ${}^{12}_{6}C$, the reaction is

$$ {}^{12}_{6}C = 6\,{}^{1}_{1}H + 6\,{}^{1}_{0}n. \qquad (2)$$

For a nucleus, the average binding energy is defined as the total binding energy divided by the value of A. The total binding energy is the energy for reaction (1) as calculated from the difference in the rest masses of products and reactants. Therefore, the average binding energy for reaction (2) is given by

$$\frac{\Delta(\text{mass})}{12} = \frac{6(1.007825) + 6(1.008665) - 12.000000}{12}$$

$$= \frac{0.098940 \text{ u}}{12}$$

lu = 1 amu, $\Delta E = (\Delta m)c^2$ where c is the speed of light. Therefore,

$$\frac{0.09894\text{amu} \times 1.6605(10^{-24})\text{g amu}^{-1} \times (2.997925(10^{10})\text{cm sec}^{-1})^2 \times 10^{-7}\text{J sec}^{-1}}{1.6022\ (10^{-19})\text{J ev}^{-1} \times 10^6 \text{ev Mev}^{-1}}$$

$= 92.163$ Mev

Thus, $$\frac{\Delta(\text{mass})}{12} = \frac{92.163 \text{ Mev}}{12}$$

$$= 7.680 \text{ Mev}$$

The packing fraction is given by

$$pf = \frac{(M - A)\ 10^4}{A} \qquad (3)$$

where pf = packing fraction. M – A = mass excess = mass of the nuclide in u less the mass number. In this problem, observe that M = A. As a result, equation (3) yields.

$$pf = \frac{(12 - 12)\ 10^4}{12}$$

$$= \frac{(0)\ 10^4}{12}$$

or $$pf = 0$$

Q. 8 : Calculate the recoil energy of an atom of mass 10, following emission of a 2 Mev gamma ray.

Solution : The energy equivalent to one mass unit is $E = mc^2 = 931 \times 10^6$ ev.

$$E_r = \frac{(2 \times 10^6)^2}{1862 \times 10^6 \times 10} = 214 \text{ ev},$$

where E_r = recoil energy

or, alternatively,

$$E_r = \frac{E_\gamma^2}{2\,Mc^2}$$

$$= \frac{4 \times 10^{12}\,(ev)^2}{2 \times 10\,\dfrac{g}{mole} \times \dfrac{mole}{6.02 \times 10^{23}\ molec} \times 9 \times 10^{20}\,\dfrac{cm^2}{sec^2}}$$

$$= \frac{4 \times 10^{12}\,(ev)^2}{\dfrac{180 \times 10^{20}\ ergs}{6 \times 10^{23}\ molecules} \times \dfrac{ev}{1.6 \times 10^{-12}\ ergs}}$$

$$= \frac{4 \times 10^{12}\ ev}{186.2 \times 10^{8}\ recoil}$$

$$E_r = 214\ ev/recoil$$

Q. 9 : Calculate ΔE for the proposed basis of an absolutely clean source of nuclear energy

$$^{11}_{5}B + {}^{1}_{1}H \rightarrow {}^{12}_{6}C \rightarrow 3\,{}^{4}_{2}He$$

Atomic masses : $^{11}B = 11.00931$, $^{4}He = 4.00260$, $^{1}H = 1.00783$.

Solution : The energy of nuclear reactions, ΔE, is calculated from the difference between the masses of products and reactants in accordance with the Einstein Law. Einstein's Law can be stated as $\Delta E = \Delta mc^2$, where Δm is the difference in the masses of the products and the reactants, and c is the speed of light (3×10^{10} cm/sec). The total reaction here can be written

$$^{11}_{5}B + {}^{1}_{1}H \rightarrow 3\,{}^{4}_{2}He.$$

Δm is equal to the mass of $^{11}_{5}B$ and $^{1}_{1}H$ substracted from the mass of 3 $^{4}_{2}He$. The mass of 3 $^{4}_{2}He$ is equal to 3 times the mass of $^{4}_{2}He$.

Thus

$$\Delta m = \left(3 \times m\ of\ {}^{4}_{2}He\right) - \left(m\ of\ {}^{11}_{5}B + m\ of\ {}^{1}_{1}H\right)$$

$$= (3 \times 4.00260) - (11.00931 + 1.00783)$$

$$= 12.0078 - 12.01714 = -9.34 \times 10^{-3}\ g/mole.$$

Now, using Δm to solve for ΔE yields.

$$\Delta m = \Delta mc^2 = -9.34 \times 10^{-3} \text{ g / mole} \times \left(3.0 \times 10^{10} \text{cm / sec}\right)^2$$

$$= -9.34 \times 10^{-3} \text{ g / mole} \times 9.0 \times 10^{20} \text{ cm}^2 / \text{sec}^2$$

$$= -8.406 \times 10^{18} \text{ g cm}^2 / \text{mole sec}^2$$

$$= -8.406 \times 10^{18} \text{ ergs / mole.}$$

There are 4.18×10^{10} ergs in 1 Kcal, thus ergs can be converted to kcal by dividing the number of ergs by the conversion factor, 4.18×10^{10} ergs / kcal.

$$\text{no. of kcal} = \frac{-8.406 \times 10^{18} \text{ ergs / mole}}{4.18 \times 10^{10} \text{ ergs / kcal}}$$

$$= -2.01 \times 10^{8} \text{ kcal / mole.}$$

ΔE for this reaction is -2.01×10^{8} kcal per mole.

Q. 10 : Calculate the yield, the threshold energy and the projectile energy for the reaction

$$^{58}_{26}\text{Fe}\left(^{3}_{2}\text{He},\ ^{2}_{1}\text{H}\right)\ ^{59}_{27}\text{Co}$$

if the masses are 57.9333, 3.01603, 2.0140 and 58.9332 u, respectively.

Solution : The yield of the reaction is defined as the negative of the energy for a nuclear transformation reaction. This energy is calculated by subtracting the sum of the rest masses of the reactants from the sum of the rest masses of the products. That is,

$$Q = -\Delta \text{ (mass)} \qquad (1)$$

where Q = the yield

D (mass) = (rest masses of products) – (rest masses of reactants).

Substitute the given data into equation (1) to obtain

$$Q = -\Delta(\text{mass}) = -[(58.9332 + 2.0140) - (57.9333 + 3.01603)]$$

$$= -[60.9472 - 60.94933]$$

$$= -[-0.00213]$$

$\therefore \quad Q = 0.0021 \text{ u} = 1.96 \text{ MeV}$

Observe that Δ (mass) is negative. Because of this, the threshold energy, which is given by

$$E_{thr} = \frac{\left(M_{proj} + M_{target}\right)\Delta(\text{mass})}{M_{target}}$$

has no meaning.

The projectile energy is given by

$$E_{proj} = \frac{\left(M_{proj} + M_{targ}\right)U_{proj}}{M_{targ}} \qquad (2)$$

where U_{proj} for ^3_2He is given as

$$U_{proj} = \frac{(1.92 \text{ MeV})\left(P_{targ}\right)}{A_{targ}^{\frac{1}{3}} + 1.44} \qquad (3)$$

P is the number of protons in the nucleus, otherwise known as the atomic number.

A = mass number or the number of protons and neutrons in the nucleus. From equation (3),

$$U_{proj} = \frac{(1.92)(26)}{58^{\frac{1}{3}} + 1.44} = 9.40 \text{ MeV}$$

Using this value of U_{proj} in equation (2) gives.

Q. 11 : Calculate the maximum kinetic energy of the β^- emitted in the radioactive decay of He^6.

Solution: To solve this problem, one must first know that β^- is an electron of negative unit charge and zero mass. Thus, the equation for this particular decay process is

$$^6_2He \rightarrow {}^6_3Li + \beta^-. \quad \text{or} \quad {}_2He^6 \rightarrow {}_3Li^6 + {}_{-1}e^0.$$

The kinetic energy of the β^- particle comes from the difference in mass (or energy, $E = mc^2$), between products and reactants. To compute the mass change during this process, only the whole atomic mass of He^6 and Li^6 need be considered. Once the difference in mass is known, a conversion factor is used (931. MeV = 1 amu). Thus

$$\begin{aligned} \text{Mass of He}^6 &= 6.01890 \text{ amu} \\ \text{Mass of Li}^6 &= 6.01513 \text{ amu} \\ \text{loss in mass} &= 0.00377 \text{ amu} \end{aligned}$$

Energy equivalent = (931 MeV/1 amu) (0.00377 amu) = 3.51 MeV.

Therefore, the maximum kinetic energy of the β^- particle is 3.51 MeV.

Q. 12 : Calculate the Coulombic barrier in the bombardment of ^{209}Bi with alpha particles.

Solution : In bombarding a nucleus with a charged particle an energy barrier is encountered. This is due to the Coulombic repulsion which develops between the charged particle and the nucleus. An empirical formula to calculate this Coulombic barrier is given by

$$E_{barrier} = \frac{0.96\, zZ}{A_{particle}^{1/3} + A_{target}^{1/3}}$$

where

$$\begin{aligned} z &= \text{charge number of the particle} \\ Z &= \text{atomic number of the target} \\ A_{particle} &= \text{mass of the particle} \\ A_{target} &= \text{mass of the target} \\ B_{barrier} &= \text{Coulombic barrier measured in MeV} \end{aligned}$$

For the bombardment of ^{209}Bi with alpha particles, z = 2, Z = 83, $A_{particle}$ = 4 and A_{target} = 209. Therefore,

$$E_{barrier} = \frac{(0.96)(2)(83)}{(4)^{1/3} + (209)^{1/3}} = 21.2 \text{ MeV}$$

15

Surface Chemistry

Q. 1 : Atoms of argon gas are being adsoı bed onto a uniform surface of a certain crystal and the adsorbed atoms are freely mobile. Calculate (a) the adsorption coefficient using the Langmuir adsorption isotherm by treating the adsorbed phase as a two-dimensional gas. (b) θ (P, T) for P = 1 atm, θ = 200°K and $\Delta\varepsilon_0$ (adsorption) = 0.35 eV.

Solution : The equation that relates the amount of gas adsorbed on a surface to the pressure of the gas at constant temperature is called the adsorption isotherm. One such isotherm is the Langmuir adsorption isotherm, given by

$$\theta = \frac{b(T)P}{1 + b(T)P} \qquad (1)$$

where θ = the fraction of the surface area covered by adsorbed molecules at any time and P is the pressure. b (T) is the adsorption coefficient and is given in terms of statistical mechanics as

$$b(T) = Q(T)\, e^{-\mu^0/kT} \qquad (2)$$

where Q = the partition function.

For a mobile two-dimensional adsorbed gas, there is free translational motion in the xy plane of the crystal surface and with a vibration of magnitude of 5×10^{12} sec^{-1} perpendicular to the surface.

The partition function of the argon gas is given by

$$Q(T) = \left(Q_{x,\,trans.}\,Q_{y,\,trans.}\,Q_{z,\,vib}\right)e^{-\Delta\varepsilon_0/kT} \tag{3}$$

where $Q_{x,\,trans}$, $Q_{y,\,trans.}$ = the translational partition function in the x- and y-plane respectively. $Q_{z,\,vib.}$ is the vibrational partition function in the z-plane,

$$Q_{x,\,trans.} = Q_{y,\,trans.} = \frac{(2\pi mkT)^{1/2}\,a}{h} \tag{4}$$

where h = Planck's constant. For two degrees of translational freedom, equation (4) is squared, and since a^2 = area, A, equation (4) changes to

$$Q_{x,\,trans.}\,Q_{y,\,trans.} = \frac{(2\pi mkT)\,A}{h^2}. \text{ A is in m}^2.$$

$$\therefore \quad Q_{x,\,trans.} = \frac{2\pi\left(40\times10^{-3}\text{ kg}\right)\left(1.380\times10^{-23}\text{ J° K}^{-1}\right)\left(200°\text{ K}\right)(A)}{\left(6.022\times10^{23}\text{ mole}^{-1}\right)\left(6.62\times10^{-34}\text{ J}-\text{sec}\right)^2}$$

$$= 2.69 \times 10^{21}\text{ A}.$$

$$Q_{z,vib.} = \frac{e^{-h\nu/2kT}}{1-e^{-h\nu/2kT}}$$

$$= \frac{e^{h\nu/2kT}}{e^{h\nu/2kT}-1}$$

$$\frac{h\nu}{2kT} = \frac{\left(6.62\times10^{-34}\text{ J}-\text{sec}\right)\left(5\times10^{12}\text{ sec}^{-1}\right)}{2\left(1.3805\times10^{-23}\text{ J° K}^{-1}\right)\left(200°\text{ K}\right)}$$

$$= 0.60.$$

Therefore,

$$Q_{z,vib.} = \frac{e^{0.60}}{e^{1.20}-1}$$

$$= 0.785$$

$$\frac{\Delta\varepsilon_0}{kT} = \frac{(0.35\text{eV})\left(1.602\times10^{-19}\text{ J/ev}\right)}{\left(1.3805\times10^{-23}\text{ J° K}^{-1}\right)\left(200°\text{ K}\right)}$$

$$= 20.31.$$

Thus $e^{-\Delta\varepsilon_0/kT} = e^{-20.31}$

$$= 1.515 \times 10^{-9}.$$

From equation (3),

$$Q(T) = \left(2.629 \times 10^{21}\ A\right)(0.785)\left(1.515 \times 10^{-9}\right)$$

$$= 3.13 \times 10^{12}\ A$$

$$\frac{\mu^0}{kT} = \frac{-kT \ln\left[\left(\frac{2\pi mkT}{h^2}\right)^{3/2} \frac{RT}{LP}\right]}{kT}$$

where L = Avogadro's number and P is the pressure. Simplification of the above equation yields

$$\frac{\mu^0}{kT} = -\ln\left[\left(\frac{2\pi mkT}{h^2}\right)^{3/2} \frac{RT}{LP}\right]$$

$$= -\ln\left\{\left[\frac{2\pi\left(40 \times 10^{-3}\ \text{kg mole}^{-1}\right)\left(1.3805 \times 10^{-23}\ \text{J}\,^\circ\text{K}^{-1}\right)\left(200^\circ\ \text{K}\right)}{\left(6.022 \times 10^{23}\ \text{mole}^{-1}\right)\left(6.62 \times 10^{-34}\ \text{J}-\text{s}\right)^2}\right]^{3/2}\right.$$

$$\left.\left[\frac{\left(0.082 \times 10^{-3}\ \text{m}^3\ \text{atm}\,^\circ\text{K}^{-1}\ \text{mol}^{-1}\right)\left(200\ ^\circ\text{K}\right)}{\left(6.022 \times 10^{23}\ \text{mol}^{-1}\right)\left(1\ \text{atm}\right)}\right]\right\}$$

$$= -\ln\left[\left(1.35 \times 10^{32}\right)\left(2.72 \times 10^{-26}\right)\right]$$

$$= -\ln\left(3.672 \times 10^{6}\right)$$

$$= -15.12$$

$$e^{\mu^0/kT} = e^{-15.12}$$

$$= 2.72 \times 10^{-7}.$$

From equation (2) the adsorption coefficient is

$$b(T) = Q(T)\, e^{\mu^0/kT}$$

$$= \left(3.13 \times 10^{12}\ A\right)\left(2.72 \times 10^{-7}\right)$$

$$= 8.51 \times 10^{5}\ A$$

where A = the area of the surface in m^2.

(b) The fraction of occupied sites on the surface of the crystal, θ, is given by equation (1). Therefore,

$$\theta = \frac{(8.51 \times 10^5\ A)\,(1\ atm)}{1 + 8.51 \times 10^5\ A}$$

Q. 2 : The data below are for the adsorption of CO on charcoal at 273K. Confirm that they fit the Langmuir isotherm, and find the constant K and the volume corresponding to complete surface coverage.

p/mmHg	100	200	300	400	500	600	700
V/cm³	10.2	18.6	25.5	31.4	36.9	41.6	46.1

Mass of sample of charcoal : 3.022 g; in each case V has been corrected to 1 atm.

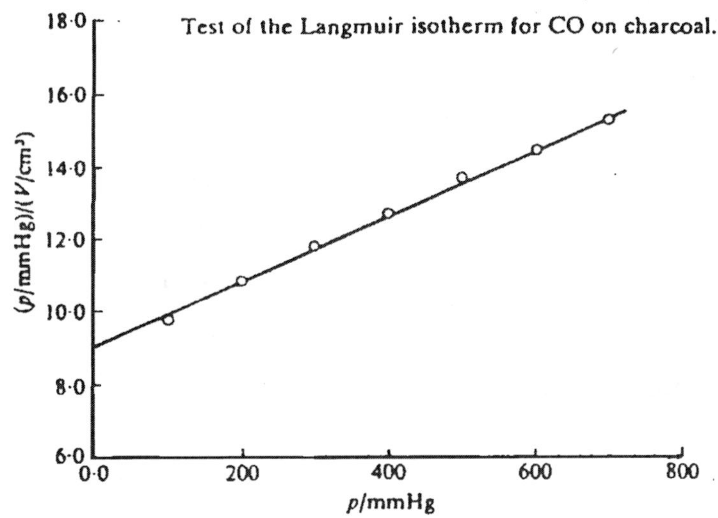

Test of the Langmuir isotherm for CO on charcoal.

Solution : The Langmuir isotherm is given by $\theta = \dfrac{Kp_A}{1 + Kp_A}$ where p_A = pressure of A, K = ratio of the rate constant for adsorption to that for desorption. That is, $K_a / K_d = K$ from the above equation,

$$Kp_A = \theta + Kp_A\theta. \qquad (1)$$

$$\theta = \frac{\text{number of adsorption sites filled}}{\text{number of adsorption sites available}}.$$

Let $\theta = V / V_\infty$ where V_∞ represents complete coverage. Then, equation (1) becomes

$$\frac{p_A}{V} = \frac{1}{KV_\infty} + \frac{p_A}{V_\infty}.$$

A plot of p_A / V versus p_A should give a straight line of slope $1 / V_\infty$ and intercept $1 / KV_\infty$.

From the given data, the following table is drawn, from which the graph will be plotted.

p	100	200	300	400	500
p_A/V	$\frac{100}{10.2} = 9.8$	$\frac{200}{18.6} = 10.8$	$\frac{300}{25.5} = 11.8$	$\frac{400}{31.4} = 12.7$	$\frac{500}{36.9} = 13.6$

600	700
$\frac{600}{41.6} = 14.4$	$\frac{700}{46.1} = 15.2$

The values are plotted below.

$$\text{Slope of the line} = \frac{15.2 - 9.8}{700 - 100} = \frac{5.4 \dfrac{1}{\text{mm Hg cm}^2}}{600 \dfrac{1}{\text{mm Hg}}}$$

$$= 0.0090 \text{ cm}^{-3}$$

Since slope $= 0.009 \text{ cm}^{-3} = 1 / V_\infty$,

$$V_\infty = \frac{1}{0.009 \text{ cm}^2} = 111 \text{ cm}^3.$$

The intercept is approximately 9.0 as shown on the graph.

$$\text{intercept} = 9.00 = \frac{1}{KV_\infty} = \frac{1}{(K)\ (111)}$$

$$K = \frac{1}{(9)\ (111)}$$

$$= 0.001 \text{ mm Hg}^{-1}.$$

Q. 3 : The data for the adsorption of oxygen on smooth iron at −183°C, as investigated by Armbruster and Austin [J. Am. Chem. Soc., 68, 1347 (1946)], is given below.

t, sec	V, cm^3
1380	0.167
3000	0.272
4260	0.330
7320	0.408
10020	0.432

where V = the volume adsorbed on a particular specimen of iron up to time t, and V_e = 0.451 cm3 is the volume adsorbed at equilibrium. Show that the adsorption rate follows first-order kinetics and determine the rate constant, k.

Solution :

t, sec	V, cm^3	V_e	$(V_e - V)$	$\ln(V_e - V)$
1380	0.167	0.451	0.284	−1.259
3000	0.272	0.451	0.179	−1.720
4260	0.330	0.451	0.121	−2.112
7320	0.408	0.451	0.043	−3.147
10020	0.432	0.451	0.019	−3.963

Plot $\ln(V_e - V)$ vs t. Obtain a straight line with a slope of − k. The value of k is estimated to be about 1.38×10^{-4} sec^{-1}.

Q. 4 : The following data was obtained for the adsorption of aqueous acetic acid on charcoal.

c_0, molarity of acetic acid in solution before addition to charcoal	c_e, molarity of acetic acid remaining in solution at equilibrium	m, grams of charcoal
0.503	0.434	3.96
0.252	0.202	3.94
0.126	0.0899	4.00
0.0628	0.0347	4.12
0.0314	0.0113	4.04
0.0157	0.0033	4.00

Show that these data is in accordance with the Freundlich adsorption isotherm,

$$\frac{x}{m} = kc_e^{1/n}$$

where x = the number of grams of acetic acid adsorbed and k and n are constants.

Estimate the constants k and n. In all cases the volume of the solution in contact with the charcoal was 200 ml.

Solution : Adsorption from solution does not in general appear to lead to layers more than one molecule thick. For many cases the experimental data can be fairly well represented by an empirical isotherm proposed by Freundich.

$$\frac{x}{m} = kc_e^{1/n}. \quad (1)$$

Here, x and m are the masses of the substances adsorbed (adsorbate) and of the substances on which the adsorption takes place (adsorbent) respectively. c_e is the concentration of the solution at equilibrium and n is an empirical constant usually greater than unity. Freundlich equation then implies that if log x/m is plotted against log c_e, a straight line will be obtained with a slope of 1/n.

Let x = number of grams adsorbate (CH_3COOH) adsorbed
= [$(c_0 - c_e)$ mole / liter] [0.200 liter] [60.0 g / mole].

From equation (1)

$$\log \frac{x}{m} = \log k + \frac{1}{n} \log c_e$$

c_0	c_e	x	$\frac{x}{m}$	$\log\left(\frac{x}{m}\right)$	$\log c_e$
0.503	0.434	0.828	0.209	–0.680	–0.362
0.252	0.202	0.600	0.152	–0.818	–0.695
0.126	0.0899	0.433	0.1083	–0.965	–1.046
0.0628	0.0347	0.337	0.0818	–1.087	–1.459
0.0314	0.0113	0.241	0.060	–1.222	–1.950
0.0157	0.0033	0.148	0.037	–1.430	–2.478

Plot $\log\left(\frac{x}{m}\right)$ vs log c_e to get a straight line. The intercept for c_e

= 1 (log c_e = 0) = log k. Therefore, log k = 9.448 – 10 = 0.552 and k = 0.28, slope = 1/n = 0.355 and n = 2.82.

Q. 5 : The surface tension, γ, of an aqueous solution of butyric acid is related at 18°C to the bulk concentration, c, by the equation

$$\gamma_0 - \gamma = 29.8 \text{ dyne/cm} \log (1 + 19.64 \text{ M}^{-1} \text{ c})$$

where γ_0 is the surface tension of pure water. Calculate the excess concentration s, of solute per square centimeter of surface when c = 0.01 M. Use Gibbs adsorption equation. Determine the limiting value of S as c becomes infinite.

Solution : $\gamma_0 - \gamma = 29.8$ dyne/cm log (1 + 19.64 M^{-1} c). To convert logarithms in base 10 to natural logarithms the following conversion is used :

$$2.303 \log x = \ln x.$$

The reason for the nesessity of this conversion is to use the readily known formula for the derivative of ln x. Hence

$$\gamma_0 - \gamma = 29.8\left(\frac{1}{2.3026}\right) \ln\left(1 + 19.64c\right).$$

Therefore,

$$-\frac{d\gamma}{dc} = (29.8) \text{ dyne/cm } (.4343)\left(\frac{1}{1+19.64c}\right)(19.64). \quad (1)$$

The Gibbs adsorption equation is given by

$$S = \frac{-c}{RT}\frac{d\gamma}{dc} \quad (2)$$

where S = the excess concentration of solution per square centimeter of surface and c is the concentration. From equation (1).

$$\frac{d\gamma}{dc} = \frac{-0.4343 \times 29.8 \text{ dyne/cm} \times 19.64 \text{ M}^{-1}}{1 + 19.64 \text{ M}^{-1} \text{ c}}$$

$$= \frac{-254 \text{ dyne cm}^{-1} \text{ M}^{-1}}{1 + 19.64 \text{ M}^{-1} \text{ c}}$$

Rearranging equation (2) solve for S yields

$$S = -\frac{c}{RT} \cdot \frac{d\gamma}{dc}$$

$$= \frac{(254c)\ \text{dyne M}^{-1}\ \text{cm}^{-1}}{RT\left[1 + \left(19.64\ \text{M}^{-1}\right)c\right]}$$

$$= \frac{254\ \text{dyne M}^{-1}\ \text{cm}^{-1} \times 0.01\ \text{M}}{\left(8.314 \times 10^{7}\ \frac{\text{erg}}{°\text{K mole}}(291\ °\text{K})\right) \times \left(1 + 19.64\ \text{M}^{-1} \times 0.01\ \text{M}\right)}$$

$$S = 8.79 \times 10^{-11}\ \text{moles cm}^{-2}$$

(b) $$S = \frac{254\ c}{RT(1 + 19.64\ c)} = \frac{254\ c}{RT + 19.64\ RTc}$$

To calculate the limit of S as c approaches infinity, divide the right side of the above equation by the highest power of c. Hence,

$$\lim_{c \to \infty} S = \frac{254\ c}{RT/c + 19.64\ RTc/c} = \frac{254}{RT/c + 19.64\ RT}.$$

Now as $c \to \infty$, $RT/c \to o$. Therefore,

$$\lim_{c \to \infty} S = \frac{254\ \text{dyne M}^{-1}\ \text{cm}^{-1}}{19.64\ RT}$$

$$= \frac{254\ \text{dyne M}^{-1}\ \text{cm}^{-1}}{\left(19.64\ \text{M}^{-1}\right)\left(8.31 \times 10^{7}\right)\frac{\text{erg}}{°\text{K mole}}(291\ °\text{K})}$$

$$= 5.34 \times 10^{-10}\ \frac{\text{moles}}{\text{cm}^2}.$$

Q. 6 : At 25°C, the mean activity coefficient $\gamma_\pm$ of KCl in aqueous solution of 0.002 molal is 0.9648. Calculate the size droplets of 0.002M KCl solution at 25°C that would be in equilibrium with a plane surface of pure water.

Solution : By definition, the activity of KCl (aq) is given by

$$a_{KCl}\ (aq) = \gamma_\pm^2\ m^2 \qquad (1)$$

where m = the molality and $\gamma_\pm$ is the mean activity coefficient. One form of the Gibbs adsorption isotherm equation is given by

$$\Gamma = -\frac{1}{kT}\frac{d\gamma}{d\ln a_{KCl}} \qquad (2)$$

where Γ = the concentration of adsorbed component of the solution at the interface, and k = Boltzmann's constant, which is the gas constant in ergs divided by Avogadro's number.

Substitute equation (1) into quation (2) to obtain

$$\Gamma = -\frac{1}{2kT}\;\frac{d\gamma}{d\ln\gamma_{\pm}\,m}$$

$$= -\frac{\gamma_{\pm}\,m}{2kT}\;\frac{d\gamma}{d(\gamma_{\pm}\,m)}$$

At low concentrations, $\frac{d\gamma}{d(\gamma_{\pm}\,m)} = \frac{d\gamma}{dm}$. Therefore,

$$\Gamma = -\frac{\gamma_{\pm}\,m}{2kT}\;\frac{d\gamma}{dm} \qquad (3)$$

At 20°C the surface tension of water = 72.75 dyne cm^{-1} and that for a 0.74 wt% KCl solution (0.100 m) = 72.99 dyne cm^{-1}.

$$\frac{dy}{dm} = \frac{0.240}{0.100}$$

$$= 2.40 \text{ dyne cm}^{-1}\text{ molal}^{-1}.$$

❑❑❑